ORGANIZATIONAL LELS

CONTEXTS OF LEARNING
Classrooms, Schools and Society

Managing Editors:

Bert Creemers, *GION, Groningen, The Netherlands*
David Reynolds, *School of Education, University of Newcastle upon Tyne,
    England*
Sam Stringfield, *Center for the Social Organization of Schools, Johns Hopkins
    University*

ORGANIZATIONAL LEARNING IN SCHOOLS

EDITED BY
KENNETH LEITHWOOD
and
KAREN SEASHORE LOUIS

Taylor & Francis
Taylor & Francis Group

LONDON AND NEW YORK

Published by Taylor & Francis
2 Park Square, Milton Park, Abingdon, Oxon, OX14 4RN
270 Madison Ave, New York NY 10016

Transferred to Digital Printing 2006

Library of Congress Cataloging-in-Publication Data
applied for

Cover design: Ivar Hamelink
Typesetting: Red Barn Publishing, Skeagh, Skibbereen, Co. Cork, Ireland

ISBN 90 265 1539 1 (hardback)
ISBN 90 265 1540 5 (paperback)

Contents

Part IV

1

Organizational Learning in Schools: An Introduction

Kenneth Leithwood and Karen Seashore Louis

> Organizational learning is changed organization capacity for doing something new. (Watkins & Marsick, 1993, p. 148)

In her 1996 address to the annual meeting of the American Educational Research Association, Linda Darling-Hammond (1996) argued that, in order for all children to learn, we:

> need to understand how to teach in ways that respond to students' diverse approaches to learning, that are structured to take advantage of students' unique starting points, and that carefully scaffold work aimed at more proficient performances. (p. 7)

The aim of such "adaptive education" is to give "students access to social understanding . . . by talking and making decisions with one another and coming to understand multiple perspectives" (p. 6). Additionally, and a crucial part of Darling-Hammond's position for purposes of this book, she also argued that we must "understand what schools must do to organize themselves to support such teaching and learning" (p. 7). Darling-Hammond is not the first to suggest that educational restructuring and reform should be undertaken from the bottom up or that the design of schools should focus around the students' experience. But, in order to do this, Darling-Hammond also explicitly called for schools to become "learning organizations." What does this mean and why should that be advocated as an image of future schools?

Organizational learning in schools

Organizations continuously accumulate experiences that either reinforce or change their behavior. What is being learned, however, may be useful or dangerous, mundane or insightful, may lead to change or provide ways of avoiding change. When the term "learning organization" is used it usually implies the most desirable of each of these pairs of alternatives associated with learning. As a number of authors have noted, organizational learning is not the same as problem-solving or decision-making: productive group learning often occurs in situations where addressing a specific need is not the focus of attention. Nor do we believe that there is a simple way of distinguishing between schools that are "learning organizations" and those that are not. However, most writers on the topic define it as:

> A group of people pursuing common purposes (and individual purposes as well) with a collective commitment to regularly weighing the value of those purposes, modifying them when that makes sense, and continuously developing more effective and efficient ways of accomplishing those purposes. (Leithwood & Aitken, 1995, p. 41)

In this book, therefore, we take the position that what matters most are learning processes in organizations and ways of enhancing the sophistication and power of these processes.

The authors of the chapters in this book each explicitly address the meaning(s) of organizational learning which they adopted for their own purposes. Given such attention throughout the book, suffice it to say, by way of introduction, our meaning of this term includes all of the following:

- the learning of individuals in organizational contexts, represented in the work of Brown and Duguid (1996) on learning in "communities of practice" for example
- the learning that occurs in small groups or teams as, for example, described in Hutchins' (1996) study of the "mutual adaptation" of team members in their efforts to solve a problem of critical proportions
- the learning that occurs across organizations as a whole, exemplified in Cook and Yanow's (1996) analysis of the adoption of a new technology in a traditional manufacturing setting. This was a cultural process in which "the dynamic, ongoing preservation of organizational identity is as compelling as an exclusive focus on learning new things and unlearning outlived ones" (p. 451)

Most chapters of the book are concerned with the second and third of these meanings of organizational learning.

The case for organizational learning

The call for organizations of all kinds to focus more on the individual and collective learning of their members is ubiquitous, and not new. With private sector organizations in mind, Edmondson and Moingeon (1996) argue, for example, that "organizations facing uncertain, changing or ambiguous market conditions need to be able to learn" (p. 7)—a sentiment that echoes a perspective that was articulated as early as the late 1960s (Terreberry, 1968). They need to learn adaptive responses to these uncertainties and ambiguities: this is learning *what to do*, what types of new practices to adopt. These organizations also need to learn *how to do* these organizational practices because, as Argyris (1996) explains:

> no managerial theory, no matter how comprehensive, is likely to cover the complexity of the context in which the implementation is occurring. There will always be gaps and there will always be gap-filling. Organizational learning is critical to detecting and filling the gaps. (p. 1)

Schools clearly qualify as organizations facing the changing, uncertain, and ambiguous conditions alluded to by Edmonson and Moingeon. The terms "reform" and "restructuring" are in imminent danger of becoming simply catch-all phrases for a host of complex and often largely untested changes hotly advocated by politicians, parents, professionals and academics in developed countries around the world. Such changes create the same conditions of instability, and impermanence faced by most organizations approaching the end of the millenium. They also severely challenge virtually all organizational designs that rely on centralized planning, control, and direction, at the same time that some critics of schools look for simple "one-size-fits-all" solutions such as improved national tests or more a more standardized curriculum.

Theorists have begun to acknowledge that such conditions require that organizations have the capacity for self organization: the complexity of this postmodern environment demands full use of the intellectual and emotional resources of organizational members (Handy, 1989; Morgan, 1986). Mitchell, Sackney, and Walker (1997) argue, "The postmodern era suggests a conception of organizations as processes and relationships rather than as structures and rules" (p. 52), with conversation as the central medium for the creation of both individual meaning and organizational change. From this perspective, the image of schools as learning organizations seems like a promising response to the continuing demands for restructuring.

We can further clarify the strengths of approaching school reform and school redesign through the development of organizational learning processes when we compare this approach with two alternatives that have received greater attention in the educational literature to date. These are the "effective schools" and "backward mapping" alternatives.

Effective schools

The best known and most fully tested of these school reform and redesign alternatives entails rebuilding schools to reinforce the statistical correlates of effectiveness which have emerged out of extensive research in a number of countries (e.g. Scheerens and Bosker, 1997; Teddlie & Stringfield, 1993). In these studies, a school's effectiveness is measured by the performance of students on standardized tests—usually those focusing on "basic skills" in reading and mathematics. There is some overlap between the most common effectiveness correlates and the conditions fostering organizational learning, for example, school culture, and collabortive decision-making structures (Cheng, 1996; Heck & Marcoulides, 1996). But there are large differences as well. For example: district conditions are only rarely considered relevant in the effective schools literature, whereas they appear to have considerable importance in fostering organizational learning (Coffin, 1997; Leithwood, Leonard & Sharratt, in press; McLaughlin, 1990); and the "instructional" models of leadership supported by the effective school research (Mortimore, 1993) are more control-oriented and less oriented toward capacity-development than the transformational leadership practices which, theoretically at least, seem likely to foster organizational learning.

More fundamental, however, is the crucial assumption implicitly embedded in much of the effective schools literature that a single best, albeit elusive, organizational design is suitable for most schools (Creemers & Reezigt, 1996). This assumption may be justifiable for schools pursuing essentially the same outcomes as those used as dependent measures in effective schools research, under at least largely similar circumstances. In contrast, an organizational learning approach to school design assumes that the initial conditions for effective learning must be established in many schools through special efforts which often may have to be launched from outside the school. But, given reasonable success through those initiatives, it is likely that subsequently the organization will not only be "self-organizing" but also "continually refining" in response to changes in its goals and the circumstances in which those goals are to be achieved. At any point in time, schools adhering to this approach could look very different from one another, except in respect to those core conditions necessary to sustain and encourage organizational learning.

Backward mapping

A second, less well-explored alternative is to design schools through a process of backward mapping, an approach advocated by a prominent group of school restructuring scholars (Peterson et al., 1996). This approach, adopting an open-systems, contingency perspective, argues that there ought to be consistent and "coherent" (Fuhrman, 1993) support for those teaching and learning activities which constitute the technical core of the school's activities. When changes are undertaken in the technical core, the rest of the school should be appropriately redesigned to be as supportive as possible to those new practices. Similar to the organizational learning approach, backward mapping offers no specific model of a well-designed school, and assumes that many designs are possible depending on the nature of the technical core and its demands. While the basic premise of the backward mapping argument appears sensible on initial examination, it gives rise to four implications that are more debatable.

First, backward mapping implies a large universe of reasonable organizational design alternatives. However, the research literature suggests there is an upper limit on qualitatively different, identifiable designs that are available. Although the number of theoretical alternatives is large, people in "real" organizations do not operate under conditions that permit them to analyze more than a few options before acting. This range is likely restricted to between four and ten although many variants within these major categories can be readily imagined (e.g. Bolman & Deal, 1991; Daft, 1989; Galbraith, 1977; Mintzberg, 1983).

A second implication of the backward mapping argument is that every time a non-trivial change is made in a school's core technology, that school's structures, culture, policies and other aspects of its organization should be revisited, if not substantially revised. But this requirement overstates the differences in design demands of many different changes in core technology—and the competing demands for change outside the core. It also overstates the contribution of school, as distinct from classroom-level variables in accounting for variation in student growth (Scheerens et al., 1989; Scheerens and Bosker, 1997). Even a whole-hearted effort by staff to introduce new mathematics standards in the classroom, for example, may not require a major rethinking of the school organization. On the other hand, the introduction of a state-mandated change in the school schedule which has limited implications for classroom activities may demand substantial redesign of the school organization, including personnel and budget policies.

Furthermore, considerable evidence supports the claim that significant changes in schools require extensive amounts of time (Fullan, 1993; Louis & Miles, 1990). Weick (1994) has argued that one reason why the school system is segmented into semi-autonomous units, such as classrooms, is so that the time-consuming aspects of system change can be avoided until absolutely necessary. But this evidence has not taken into account the added demands for change accompanying the backward mapping strategy. If such a strategy were followed with every non-trivial change in core technology, schools would evolve at an even more ponderous rate of speed than is presently the case.

A third problematic implication of the backward mapping strategy is that only the demands of the core technology need attention in the creation of a coherent organizational design. Schools are not merely technical organizations. They have significant features that are institutionalized—either legally or informally—in response to local, national and international expectations regarding how schools should operate (Meyer & Rowan, 1991; Powell & DiMaggio, 1991). Thus, the assumption that schools could be designed around changes in the technical core is naïve, even if backward mapping advocates might wish to position schools more fully in the technical sector, eventually. However, virtually all well-established organizations have both a technical and an institutionalized component. So, as Mintzberg (1983) argues, virtually all organizations

must attend both to internal consistency (designed according to the requirements of their core technology) and external consistency (designed to meet the
demands of the wider polity and society).

A final dilemma for the backward mapping perspective lies in its assumption
that changes in the technical core are both scientifically based and appropriate.
We are still a long way from providing practitioners with real technical solutions, although educational research related to effective teaching and school
organization has made remarkable strides in the past few decades. Thus, the
notion of redesigning whole-school functions based on changes in the technical
core that are still, at best, imprecise seems misplaced. Could we not, for example, argue just as well that schools should be redesigned to reflect the equally
impressive increases in our understanding of the social structures that are related to resilience in children and youth?

Where next?

The difficulties inherent in the effective schools and backward mapping strategies for school restructuring, however, do not obviate the basic problem: existing school structures do not consistently support effective teaching and learning.
An adequate response to these difficulties begins by acknowledging that the task
is not just to create a school organization capable of implementing the current
set of reform initiatives such as new curriculum frameworks, teaching for understanding, and cooperative learning strategies, in the context of today's turbulent
environments. Rather, the task is to design an organization capable of productively responding, not only to such current initiatives in today's environment,
but the endless number of initiatives, including new definitions of school effectiveness, that inevitably will follow. Complexity theory (Waldorp, 1992) has
begun to influence our understanding of social organizations. The claim that
many critical social structures, like schools, are "chaotic" open social systems,
and not amenable to rational centralized management, is increasingly awarded
more respect. To the degree that schools are able to gain the capacity for self-
organizing and to make minimal demands on central control (Caine & Caine,
1997; Morgan, 1986) they will become better able to function and perform as
learning organizations, while meeting social needs.

Problems addressed by the book

There is a strong, logical case for greater attention to individual and collective
learning in schools experiencing turbulent times, but three problems remain: the
context problem, the *evidence problem,* and the *strategies problem.* This book
is designed explicitly to address these three problems.

The context problem

Virtually none of the substantial literature about the nature, causes, and consequences of organizational learning has been written with schools in mind. As a

result, although schools face the same sorts of conditions giving rise to calls for greater individual and collective learning in other types of organizations, we cannot rely on the exisiting literature to acknowledge the unique conditions and requirements shaping the learning "styles" (DiBella et al., 1996) of schools. This is the context problem, and it is partly addressed in this book by describing what organizational learning looks like in the many case schools which appear in seven of the thirteen chapters. This problem also is addressed in the majority of chapters which take as their theoretical starting points ideas about organizational learning developed in non-school contexts and explicitly adapt them for use in schools.

The evidence problem

The rational or logical case for organizational learning is compelling, indeed. But empirical support for the claim that increases in such learning will contribute to organizational effectiveness or productivity is embarrassingly slim. This is the evidence problem. Although undeniably rich conceptually, the closest most of the organizational learning literature comes to providing empirical evidence of its claims are case illustrations. Edmonson and Moingeon (1996) make this point, for example, about the assumption found in popular management literature that organizational learning is a source of competitive advantage. And Weick and Westley (1996) recently have pointed out that "there appear to be more reviews of organizational learning than there is substance to review" (p. 440). This is a comment on the state-of-the-art of empirical research on organizational learning across all organization types. A review of empirical research on organizational learning in schools alone would make a very quick read indeed.

So, while there are compelling reasons to view schools from an organizational learning perspective, and some powerful theoretical tools to shape such a perspective, empirical evidence supporting the value of this perspective is thin, to say the least. To warrant the sort of sustained and intensive attention to organizational redesign required to significantly enhance the individual and collective learning of school professionals, a more substantial body of evidence concerning the outcomes that can be expected is needed. Empirical evidence has a way of modifying what people consider to be "logical". Eight chapters in the book provide original data about some aspect of the nature, causes and consequences of organizational learning in schools.

The strategies problem

The critical nature of the evidence problem notwithstanding, action to enhance organizational learning cannot wait until the empirical case for its effects is fully made. The world of practice just does not stand still and wait for such evidence to appear. Rather, the relationship between practice and research about practice is better conceived of as parallel and symbiotic. So the strategies problem coexists with the unresolved evidence problem. In reference to the strategies problem Edmonson and Moingeon (1996) explain that while "few managers or organizational scholars would disagree with [the need for more organizational learning] most are aware of the difficulty of taking action based on this prescription" (p. 7). The development of well-tested strategies for this purpose is urgently needed if real progress is to be made in fostering organizational learning in schools and other types of organizations. Each of the four chapters in Part II of the book describes a different strategy for enhancing organizational learning in schools and offers systematic evidence concerning the impact of the strategy.

In sum, to bring an organizational learning perspective to life in schools—to give it a practical face—systematic efforts are needed which: examine the meaning of OL in schools, identify the organizational conditions which foster and inhibit it, develop specific strategies capable of moving schools in this direction, and further clarify concepts basic to our understanding of OL. It is such efforts that are reported in the four sections of this book.

Overview of the book

Part I

Chapters 2, 3, and 4, comprising the first section, present original, qualitative data about the nature of organizational learning in schools, the conditions which foster or inhibit such learning, and some of the outcomes of such learning.

Historically, one of the most productive approaches to understanding the nature of collective learning has been to study it guided by concepts developed to explain individual learning. Karen Seashore Louis and Sharon Kruse (Chapter 2) adopt this cognitive perspective on organizational learning in their study of one elementary and one secondary school. Learning of staffs in both schools is described in terms of elements of organizational memory, and information distribution and interpretation. Evidence from these schools suggests, for example, that unlearning the past can be a significant challenge to schools engaged in significant restructuring. Learning seems to occur most productively when staffs use both internal and external sources of information, when important assumptions underlying the work of the school are given ongoing attention, when teachers' learning occurs in systematic ways, and when teachers think about their roles in new ways. The intellectual vision of the principal also proved to be a powerful stimulus for collective learning. This chapter provides important lessons for the development of schools' organizational learning capacities.

In the late 1960s, Philip Jackson (1968) published his now classical study pointing to the professional isolation of teachers in their schools. He also described the effects of this isolation on their view of teaching as a profession and the sources of their continuous professional growth. Jackson's results have gone largely unchallenged over the past three decades and it is relatively common for those doing research on teaching and teachers' lives in schools today to assume that, in typical schools at least, not much has changed. But Miriam Ben Peretz and Shifra Schonmann (Chapter 3) are not among those willing to accept that assumption. In one of the more insightful studies about such matters, the authors systematically peer into the ubiquitous teachers' lounge. Attention is paid to the sorts of norms which govern teachers' behavior while in the lounge, the effects these norms have on social interactions in the lounge, and how those interactions sometimes serve as powerful sources of both individual and collective professional learning. Based on the results of this study, it is no longer safe to assume, as Jackson claimed, that teachers live their professional lives in schools isolated from one another, if they ever did.

Like Ben Peretz and Schonmann, Kenneth Leithwood, Doris Jantzi, and Rosanne Steinbach inquire in Chapter 4 about the conditions which foster and inhibit organizational learning in schools. This study, based on interviews with 72 teachers in six schools, was guided by a framework which focused attention on factors that stimulated organizational learning, organizational learning processes themselves, as teachers perceived them, and on the outcomes of such learning. The framework for this study also directed attention to the influence on organizational learning of a large number of characteristics, processes and conditions found in the school itself (e.g. culture, structure), and the wider environment in which schools are located (e.g. district, local community). Special attention was devoted to forms of leadership which support the individual and collective learning efforts of teachers. District initiatives, school culture, and transformational leadership practices on the part of principals are among the more influential factors fostering organizational learning according to the results of this study.

Part II

Chapters 5 through 8 describe four distinct strategies for developing the organizational learning capacities of schools—cognitive mapping, participatory evaluation, professional development schools, and action research. While these are by no means the only strategies for building organizational learning capacity, they are clearly promising and have been the object of some research.

Each of these strategies depends on some form of school—university partnership. And each chapter provides both a theoretical account of its focal strategy and plausible effects, along with one or more case illustrations of the strategy in use. There is likely sufficient detail about each of these strategies to permit readers to experiment with them.

Cognitive mapping, the focal strategy in Chapter 5 by Jean King, Jeffrey Allen, and Khahm Nguyen, entails the creation of networks or maps of ideas about a person's understanding of (in this case) school goals and change processes starting with information collected through a detailed interview. This chapter describes in considerable detail the authors' experience in developing and using such maps with sub-sets of teachers and administrators in one medium-sized secondary school (1,000 students). In this case, the mapping process was a means of assisting the school to better clarify its own collective understanding of the challenges it faced, the goals it should be pursuing, and the processes that would be effective for accomplishing those goals. Mapping was considered a promising strategy for these purposes through the dialogue and discussion it fostered among staff and administration. The authors draw several important lessons from this experience, lessons about the feasibility of the mapping process, refinements of the process for future use, and additional research needed to clarify the effects of cognitive mapping on organizational learning in schools.

In Chapter 6, Brad Cousins describes a strategy for enhancing organizational learning in schools referred to as "participatory evaluation". Extensive involvement in the design, conduct, and reporting of evidence concerning some aspect of work in a school by representatives of all stakeholders in that work, with the assistance of an evaluation specialist, is the meaning of participatory evaluation. Its stimulation of deep social processing of information is the rationale for proposing this strategy, a rationale similar to the one developed by authors of earlier chapters. General features of this strategy are outlined and a case study of participatory evaluation in a large secondary school, and its consequences for multiple stakeholder groups is described. Evidence reported in this chapter suggests that participatory evaluation has considerable potential for fostering collective learning in both the short and long terms. But it is also a complex strategy that can easily have unexpected consequences.

In Chapter 7, Linda Darling-Hammond, Velma Cobb, and Marcella Bullmaster trace some of the possibilities for new forms of teacher learning and leadership that permeate teaching and are accessible to all teachers who engage in the broader professional roles available in professional development schools (PDS). Such schools consist of collaborations between schools and universities that have been created to support the learning of prospective and experienced teachers while simultaneously restructuring schools and schools of education. In the course of co-constructing learning environments where novices can learn from expert practitioners, the more highly-developed professional development schools allow veteran teachers to assume new roles as mentors, university adjuncts, school restructurers, and teacher leaders. They also allow school and university educators to engage jointly in research and rethinking of practice, thus creating an opportunity for the profession to expand its knowledge base by putting research into practice, and practice into research. And they socialize entering teachers to a new kind of professional teaching role, one grounded in collaboration, critical inquiry, and a conception of teacher as a decisionmaker and designer of practice (Darling-Hammond, 1996). In these ways, professional development schools can both enable teacher leadership for the teachers who work in them and help to build a future teaching force that assumes leadership naturally as part of a more professional conception of teaching.

The authors examine the potentials of PDSs for fostering individual and collective professional learning using data from in-depth case studies of seven PDSs that are among the more mature of these supplemented by research in a number of other professional development schools where similar patterns of teacher leadership have been observed.

Coral Mitchell and Larry Sackney describe, in Chapter 8, what they discovered from using action research in one elementary school as a means of fostering organizational learning. Six different research activities were pursued over a period of six months: theoretical information provision, individual interviews, large-group reflection meetings, interaction observations, verbatim transcripts, and data summaries. Analyses of these data confirmed the importance to the staff's learning of the cognitive processes of reflection and professional conversation (in-depth discussions among staff), as well as two equally-important affective processes, invitation (making overt efforts to draw other colleagues into discussion), and affirmation (overtly valuing the contribution of one's colleagues). These processes appeared to unfold in three distinct, but interrelated phases. In the first phase, "naming and framing," discussions were characterized by description, storytelling, and suggestion; in the second phase, "analyzing and integrating," by analysis and evaluation of current practices; and in the third phase, "applying and experimenting," discussions revolved around implementation plans.

Part III

Chapters 9, 10, and 11 examine some ideas basic to the concept of organizational learning and school improvement and offer a theoretical account of the relationship between organizational learning and the development of schools as mature learning organizations.

Organizational learning, as we pointed out above, occurs at three levels: the individual, small groups or teams of people, and whole organizations. Teams are becoming a pervasive locus of work in many different types of organizations, not least restructuring schools, and team learning is an important instance of organizational learning. Using team learning as the context, Chapter 9 (by Kenneth Leithwood) takes up two fundamental problems that challenge the credibility of organizational learning as a useful organizational concept: whether it is reasonable to speak of a collective "mind" and, if it is, what is the nature of the collective learning processes that nourish such a mind. This chapter also identifies a set of conditions likely to influence team learning, as distinct from either whole organizational learning or individual learning in organizations. These conditions are placed in a framework that aims to aid reflective practice and offer some guidance for future research on team-learning processes in schools.

The perspective on team learning developed in the chapter locates the collective team mind in the patterns of action undertaken by the team as a whole. Collective team learning entails change in these patterns of action through processes of mutual adaptation. These are processes in which individual members adapt their contributions to the team partly in response to their understanding of the nature of the new challenge facing the team and partly in response to the responses of their fellow team members' actions. In this way the team's learning has the potential to both precede and contribute to the individual member's learning.

While only recently applied to schools, organizational learning perspectives have long been used to better understand many other types of organizations. This has resulted in a rich background of theoretical concepts concerning both the meaning of OL and the nature of OL processes in such organizations.

Although less well developed, a body of empirical data exploring these concepts also has emerged. In Chapter 10, Brad Cousins provides a synopsis of some of the more central of these concepts and empirical results: these are concepts and results concerning whole organization learning, for the most part, and shaped largely by cognitive approaches to such learning. Cousins reviews what is known about how organizations encode and represent knowledge as well as how they store, retrieve, and communicate such knowledge. He also examines literature about organizational knowledge development, and the mechanisms that lead to "dysfunctional learning habits." In the final section of this chapter, the reader is offered eight guidelines or implications for action, derived from the literature, for enhancing the learning capacities of schools, in particular.

In Chapter 11, Janna Voogt, Nijs Lagerweij, and Karen Seashore Louis make a distinctive contribution to understanding the relationship between educational change and organizational learning processes. Built on the concept of "school development," the authors cite evidence that planned change processes (they discuss three orientations to such processes) account for only about 15% of the changes likely to be experienced in schools. The remaining 85% are a consequence of the normal evolution of the organization (such as is suggested in the literature on organizational life cycles), and "normal crises," a term coined by Louis and Miles (1990).

The stages through which a school evolves, argue the authors, need to be understood as a product of all three sources of change. Furthermore, these changes come about partly as a result of organizational learning processes, three different types of which are examined: single, double, and triple loop learning. The chapter offers a framework showing the ways in which different approaches to planned change and different types of organizational learning are associated with internal vs. external orientations and control oriented vs. flexible orientations to change.

Part IV

The final two chapters of the book begin to identify new directions for practice and research based on what has been learned about the nature, causes, and consequences of organizational learning in earlier chapters of the book.

Sam Stringfield, in Chapter 12, uses Senge's five disciplines as a framework for uncovering the inadequacies associated with many so-called "innovations" in education. He also points out the overwhelming cost of local, school-by-school change efforts. In light of both these matters, Stringfield argues for the widespread implementation in schools of carefully tested innovations, "reforms" which, when energetically implemented, demonstrably contribute to student learning. He also argues for schools to approach their work as "high reliability organizations," the features of which he summarizes towards the end of his chapter.

In Chapter 13, we introduce the concept of professional learning communities, argue for the value of such a concept for future schools, and point to some of the work required to further develop that concept.

References

Argyris, C. (1996). Prologue: Toward a comprehensive theory of management. In B. Moingeon & A. Edmonson (eds.), *Organizational learning and competitive advantage* (pp. 1–6). London: Sage Publications.

Bolman, L.G. & Deal, T.E. (1991). *Reframing organizations*. San Francisco: Jossey-Bass.

Brown, J.S. & Duguid, P. (1996). Organizational learning and communities-of-practice: Toward a unified view of working, learning, and innovation. In M.D. Cohen & L.S. Sproull (eds.), *Organizational learning* (pp. 58–82). Thousand Oaks, CA: Sage Publications.

Caine, R.N. & Caine, G. (1997). *Education on the edge of possibility*. Alexandria, VA: ASCD.

Cheng, Y.C. (1996). *School effectiveness and school-based management: A mechanism for development*. London: Falmer Press.

Coffin, G. (1997). *The impact of district conditions on principals' experientially acquired learning*. University of Toronto, unpublished doctoral thesis.

Cook, S.D.N. & Yanow, D. (1996). Culture and organizational learning. In M.D. Cohen & L.S. Sproull (eds.), *Organizational learning* (pp. 430–459). Thousand Oaks, CA: Sage Publications.

Creemers, B. & Reezigt, G.J. (1996). School level conditions affecting the effectiveness of instruction. *School Effectiveness and School Improvement* 7 (3): 197–228.

Daft, R.L. (1989). *Organization theory and design*. Third edition. St. Paul, MN: West Publishing Company.

Darling-Hammond, L. (1996). What matters most: A competent teacher for every child. *Phi Delta Kappan* 78 (3): 193–200.

DiBella, A., Nevis, E. & Gould, J. (1996). Organizational learning style as a core capability. In B. Moingeon & A. Edmonson (eds.), *Organizational learning and competitive advantage* (pp. 38–57). London: Sage Publications.

Edmonson, A. & Moingeon, B. (1996). When to learn how and when to learn why: Appropriate organizational learning processes as a source of competitive advantage. In B. Moingeon & A. Edmonson (eds.), *Organizational learning and competitive advantage* (pp. 17–37). London: Sage Publications.

Fuhrman, S.H. (ed.) (1993). *Designing coherent education policy: improving the system*. San Francisco: Jossey-Bass.

Fullan, M. (1993). *Change forces*. London: Falmer Press.

Galbraith, J.R. (1977). *Organization design*. Reading, MA: Addison-Wesley.

Handy, C. (1989). *The age of unreason*. Cambridge: Harvard Business School Press.

Heck, R. & Marcoulides, G. (1996). School culture and performance: Testing the invariance of an organizational model. *School Effectiveness and School Improvement* 7 (1): 76–95.

Hunter, D. (1996). Chaos theory and educational administration: Imaginative foil or useful framework? *Journal of Educational Administration and Foundations* 11 (2): 9–34.

Hutchins, E. (1996). Organizing work by adaptation. In M.D. Cohen & L.S. Sproull (eds.), *Organizational learning* (pp. 20–57). Thousand Oaks, CA: Sage Publications.

Jackson, P. (1968). *Life in classrooms*. New York: Holt, Rinehart & Winston.

Leithwood, K. & Aitken, R. (1995). *Making schools smarter*. Thousand Oaks, CA: Corwin.

Leithwood, K., Leonard & Sharratt, L. (in press). Conditions fostering organizational learning in schools. *Educational Administration Quarterly*.

Louis, K.S. & Miles, M. (1990). *Improving the urban high school: what works and why*. New York: Teachers College Press.

McLaughlin, M.W. (1990). *District contexts for teachers and teaching*. Paper presented at the annual meeting of the American Educational Research Association, Boston.

Meyer, J.W. & Rowan, B. (1991). Institutionalized organizations: Formal structure as myth and ceremony. In W.W. Powell & P.J. DiMaggio (eds.), *The New Institutionalism in organizational analysis* (pp. 41–62). Chicago, IL: University of Chicago Press.

Mintzberg, H. (1983). *Structure in fives: Designing effective organizations*. Englewood Cliffs, NJ: Prentice-Hall.

Mitchell, C., Sackney, L. & Walker (1997). The postmodern phenomena: Ramifications for school organizations and educational leadership. *Journal of Educational Administration and Foundations* 1 (11): 38–67.

Morgan, G. (1986). *Images of organization*. Beverly Hills, CA: Sage Publications.

Mortimore, P. (1993). School effectiveness and the management of effective learning and teaching. *School Effectiveness and School Improvement* 4(4): 290–310.

Peterson, P.L., McCarthy, S.J. & Elmore, R.F. (1996). Learning from school restructuring. *American Educational Research Journal* 33(1): 119–153.

Powell, W.W. & DiMaggio, P.J. (eds.) (1991). *The new institutionalism in organizational analysis*. Chicago, IL: University of Chicago Press.

Scheerens, J. & Bosker, R. (1997). The foundations of educational effectiveness. New York: Pergamon.

Scheerens, J., Vermuelen, C.J. & Pelgrum, W.J. (1989). Generalizability of instructional and school effectiveness indicators across nations. *International Journal of Educational Research* 13: 789–799.

Teddlie, C. & Stringfield, S. (1993). *Schools make a difference*. New York, NY: Teachers College Press.

Terreberry, S. (1968). The evolution of organizational environments. *Administrative Science Quarterly* 12: 590–613.

Waldorp, M.M. (1992). *Complexity: The emerging science at the edge of order and chaos*. Toronto: Touchstone.

Watkins, K. & Marsick, V. (1993). *Sculpting the learning organization*. San Francisco: Jossey Bass.

Weick, K. & Westley, F. (1996). Organizational learning: Affirming an oxymoron. In S. Clegg, C. Hardy & W. Nord (eds.), *Handbook of organization studies* (pp. 440–458). Thousand Oaks, CA: Sage Publications.

Part I

Organizational Learning in Schools: Images and Effects

2

Creating Community in Reform: Images of Organizational Learning in Inner-City Schools[1]

Karen Seashore Louis and Sharon D. Kruse

There's really not much happening ... The ones who have rushed to the apex of national attention ... they're in the business of getting funds, and staffing with directors and executive directors, and rubbing elbows with the politicians in order to have continued funding for all these efforts. I'm not naïve—how close are they to children and classrooms? You not going to change American education unless you change American teaching.

Introduction

The current reform movement is largely focused on structural and curricular changes as the main features of effective schools. Much attention is paid by politicians and national professional associations to macro-system alterations, such as increasing school autonomy through charter schools, national curriculum standards and testing. Increasingly popular local structural reforms include

1 The data collection at the two schools was supported by a grant from the Office of Educational Research and Improvement (OERI No. R117Q000005–94.). We acknowledge the efforts of our colleagues Kent Peterson, Bruce King, Donna Harris, John Balwit, and Virginia Long in conducting the interviews on which this chapter is based. The authors are solely responsible for the case studies, which are based on coded transcriptions of individual interviews in the two schools.

altering structural rigidities through block scheduling or multi-age classrooms, and changes in pedagogy, ranging from cooperative education in all subjects to problem-based science and mathematics. While recent attention is directed toward professionalizing teacher preparation and continuing licensure in a number of countries, national reform discussions in North America and Europe rarely emphasize the potential of altering the daily work-life of teachers as a means of improving both educational processes and student achievement.

However, if schools are viewed as *organizations capable of learning*—whether learning is directed toward new instructional models focused on student learning, or the development of local accountability systems—teachers' work becomes a key instrument of reform. A learning organization does not change because of directives from above, or because of alterations in the political or social environment. Rather, change and improvement occur because the individuals and the groups inside the school are able to acquire, analyze, understand and plan around information that arises from the environment and from internal monitoring. Emphasizing the school as a complex social system, rather than as a collection of structures and procedures, can help to focus our attention on the heart of the school—the teaching and learning process.

Building on an organization's ability to learn is a key element in discussions of innovation in both schools and businesses (Daft & Huber, 1987; Hedberg, 1981; Senge, 1990). Organizational learning as a model for school reform suggests that staff working within a school setting share a common social understanding related to the purposes of their work. The image of a learning organization evokes assumptions about the members of the school organization as participative, intrinsically motivated and engaged in learning with greater personal effort than other organizational models. Learning is often viewed as synonymous with individual effort and development. Personal mastery clearly has an important role to play both in the development of the individual, but also as a contributor toward any organization's effectiveness (Senge, 1990). In an organizational context, however, learning is also more than an individual response, and is focused on school-wide goals and collective effort in gathering and acting on information about performance. Information about "what we are" and "how we are doing," which in previous times might have in previous times been held by one or a few members of the school, becomes common knowledge to be acted on as a group.

Learning is also interpretive (Huber, 1991). Schoolwork is not uniform: even where there are agreed upon standards and curricula, teachers must interpret and apply what they know and agree upon in the context of specific groups of students with unique needs and abilities. In order to act collectively on what is known, school systems must include time for teachers to meet and plan lessons, units, or whole courses that combine the knowledge of the teachers who work in the particular setting. The involved teachers may or may not be members of an assigned team—schools that demonstrate high levels of organizational learning do not fit neatly into any one structural form. However, the definitions of organizational learning all share the ability to take in new information, process it, and share it among members, using regular and well-understood vehicles. What the "learning schools" that we have visited share is an inventory of prior knowledge about their school, its curriculum instructional methodology and students. In short, they are schools that "know themselves" and take the time to develop a shared vision and vocabulary with which to discuss issues of teaching and learning.

In this chapter we provide concrete illustrations of what organizational learning can look like in schools through two case studies of schools located in urban centers. Agassiz Elementary and Okanagon Middle School[2] were chosen to illustrate our basic argument: any school can become a learning organization. Neither school is particularly advantaged financially and both serve a mixed population of students, many of whom are poor. Neither would "score" at the top of their state's achievement testing programs; both, however, do far better than would be expected given their student populations. Both are considered desirable schools in which to teach, but they are not schools that had a "hand-picked" population of exceptional teachers when they began their journey into school reform. In sum, they are not "typical schools" but they do represent realistic models for development in urban settings.

After the two case studies we will introduce a model and an analysis of how organizational learning can be supported, emphasizing both the structural and cultural features of the organizational setting that support learning and development. We believe that there are many paths to becoming a learning organization. Our discussion will, therefore, focus on common features that may account for the schools' success in becoming learning organizations, but will also point to differences between the schools.

Agassiz elementary

Agassiz in context

Agassiz Elementary School is located at the edges of a sprawling school district in a southern U.S. city. From 1981 to 1988, the city's school system suffered from the chaotic effects of court-ordered desegregation that was imposed without significant assertive effort to make it work by local educational and political leaders. In 1988, after six years of "white flight," a new superintendent began a major effort to revitalize the system, using a combination of open-

---

2   The schools described here were selected from a sample of 24 who were engaged in restructuring projects (see Berends & King, 1994; Louis & Kruse, 1995, for a through discussion of the sample, data collection, and analysis methods). Our choice of these schools was determined because they were rated as high on both teacher and principal intellectual leadership by the site visit teams, and because they provided examples of existing schools with "normal accountability" that is more like other U.S. schools than some of the others that were so rated. The names of the schools and some of the features of the schools have been changed to preserve anonymity.

enrollment magnet programs, school-based management, and selective relief from highly prescriptive state curricula, personnel and testing regulations. According to one district administrator, Agassiz was one of the first and most enthusiastic participants in the new revitalization effort.

> To me [it] is really a showplace of true entrepreneurship, site-based management, parent involvement, locational budgeting, children involvement. It's just the epitome of the direction we'd like to see all schools go into . . . . some schools can't get out of the paradigm . . . Agassiz did . . . this school is creative . . . active.

Agassiz, which is located in a white, residentially segregated neighborhood, is successful in attracting African-American children from other parts of the city because of its reputation as a school that is innovative, but warm, caring, and at the cutting edge, particularly in the arts (its official magnet program) and technology. Both white and African-American teachers and parents view it as a school that provides an excellent education for children of all races, and the quality of instruction, particularly in mathematics and reading, is high.[3] For an elementary school, Agassiz is relatively large, enrolling approximately 650 students and employing 39 full-time teachers, two full-time administrators, a guidance counselor and a part-time librarian. All but one of the professional staff is female; 75% are white. One third of the students are poor, receiving federally subsidized lunch.

Key features of Agassiz: Learning from each other

Agassiz, like many forward-looking schools, has a broad array of innovative initiatives, ranging from an active Advisory Council composed of teachers and parents to special programs designed to increase student learning, particularly among students who are less well prepared. These include Reading Recovery, grade level teams to coordinate curriculum at each level, flexible staffing that has permitted teachers to make the decision to have larger classes with aides rather than more regular classroom teachers, intensive teacher involvement in hiring, and team teaching by special education and regular classroom teachers. But perhaps one of the most immediately obvious features of Agassiz is the level of enthusiasm expressed by teachers about their work—and the degree to which they view the school as more demanding than would be required of them in other schools in the district. The principal, Mrs. Cole, said:

> I think our pressures are internal . . . we could do half of what we're doing and be successful . . . but it's not enough for us.

Other teachers voice the sense of pressure and support that is felt from peers:

> There is a "healthy competitiveness" in the school . . . [but] here we are all involved . . . growing as a unit . . . it takes a team effort.

---

3 It also has a state-mandated separate program for gifted and talented students, which remains largely white due to concerns of African-American parents about cultural sensitivity and the lack of African-American teachers

> Nobody is just pulling a paycheck . . . teachers in this school believe
> [that you should] do your best . . . learn new ideas . . . and seek out-
> side views.

Many teachers also mentioned that the sense of being constantly driven to
improve meant that not all of their peers would be happy at Agassiz.

> [One teacher] tried it here for two years. She sort of didn't really give
> herself one hundred percent. Every teacher here gives one hundred
> percent. When she left she said "I guess . . . if you're mediocre, you
> don't belong at Agassiz." And I think that was right.

In addition, teachers pointed to other more subtle features of Agassiz that differentiate it from more "typical" schools in the district—even others which
are also actively involved in the district's reform program. Teachers spend a
lot of time talking to each other, and helping each other to interpret both ideas
and their own instructional practice. Much of this time is spent on Saturdays,
or after school, because the schedule does not allow for extensive meeting time
during the regular day. One teacher who was involved in the school's Reading
Recovery effort reported how powerful these interpretive experiences could
be:

> What you're talking about is a meeting that starts at 4:15 and fin-
> ishes at 7:15, so it's a long meeting . . . [we observe each other teach-
> ing through a one-way mirror]. I remember the first time I went,
> because everybody was afraid to hurt everybody's feelings . . . I'll
> never forget when Greta came out of there and said, "Alright, pan
> me. Tell me so I can grow." She was the one who turned the tide. So
> literally, we are all at a level, a professional level, where yes, we're
> embarrassed when we know we could have done things better
> behind the glass, but we're all at a level where we come out of there
> saying, "Help me, how can I help this child.?"

Teachers have also commented on the school's focus on mutual planning for
improvement:

> [We have a faculty study group] for grades K–2. There are a variety
> of opinions among those in the group. This discussion provides [us
> with] growth and exposure to other people's ideas. Most of the
> teachers are open to multiple ways of teaching a child. It also allows
> us to expand [our] ideas. It pushes [us] to think that [our own] way
> is not the only way to approach things.

One teacher emphasized that Agassiz's environment was totally different from
a previous school that she had been in, in which differences of opinion were
viewed as signs of hostility. Dialogue among teachers has recently moved beyond
discussions of coordinated planning for curriculum, to more fundamental topics related to assumptions about teaching and learning:

> We need to talk about our educational philosophy . . . we didn't have
> a common language . . . We have started to look at the book
> *Invitations* one chapter at a time and discuss it . . . we've got to read,
> reflect, think . . . gain ownership in the concepts."

She notes that the deeper discussions that are now occurring can happen
because of the conversations that have been characteristic of the school for
several years: "Dialoguing [has been] going on at grade level meetings and . . .
The reading is all [we] need to get started [on more fundamental discussions]."

In sum, Agassiz is a school where teachers regard themselves as fundamentally analytic and intellectual in regard to their work, and as part of a cohesive
team whose obligation is to help one another move forward.

Intellectual leadership from the principal

Everyone both inside and outside of the school agrees that much of the credit
for Agassiz's steady devotion to the improvement of education for all students
is the result of the principal's efforts. According to Mrs. Cole herself, a key to
her work is to provide opportunities for open discussion in "safe" conditions
where people feel free to take risks:

> Our trust level has grown . . . we used to say no to new ideas . . . due
> to policies . . . but now we actively try to do things even if initially
> policies tell us not to. We've gotten comfortable with sharing things
> that don't work . . . that have not been successful.

Teachers concur with this observation:

> I think the thing about this school is that everybody wants every-
> body to succeed. And our principal wants us to succeed. If I make
> a mistake she would never fuss, she would just say let's see if we can
> make it better. And she gives us so many opportunities to make it
> better.

In addition, Cole prides herself on being up-to-date on current issues in education, and considers herself an expert in reading (although she admitted that she
had learned a lot from teachers in the building). Her expertise is readily available
to teachers, largely through informal communication. One teacher discusses the
school norm that all teachers should read professional materials, noting that:

> [Mrs. Cole] provides educational leadership. We will be talking on a
> subject . . . and she will suggest a reading.

By promoting teacher reading, Cole stimulates teachers by creating flexible
expectations about what needs to be done. Reminding teachers that the
improvement process is a permanent feature of good schools, and that "essential questions" are more important than trivial ones, she guides teachers to critical discussion:

> [At one meeting teachers asked] "how long will this take?" She [Mrs.
> Cole] said "How long do we have? What are we doing, why are we

doing it, what can we do to make it better?" At another time after a committee meeting, Mrs. Cole asked "What do we want our graduates to look like?"

Part of Cole's style, which she describes as facilitative, is to express her own opinion, but not to impose it on others. Furthermore, many teachers describe their sense that the school fosters the norm of continuous discussion, but does not demand closure or conformity:

> I think in some schools the principal might say "this is what we're gonna do" . . . It's not done that way in this school. We have the freedom and trust from the principal that allow you to do whichever you choose.

Of course there are clear limits to this freedom. Teachers pointed out that on some key policy issues Cole listens to all voices, but makes the final decision by herself. Thus, although there is a strong belief that teachers are empowered and have an active voice in the school, Cole is ultimately in charge. Many of her "solo" decisions appear to involve disagreements among staff that need adjudication. All but one of the teachers view her exercise of the decision-making prerogative as fair.

Intellectual leadership from teachers

While the principal is behind the scenes stimulating discussion, the teachers view themselves as the most active intellectual leaders in the school. Every teacher points to some leadership role that they play in promoting discussion about critical educational ideas, either internally or with teachers outside the school. Teachers are used to sharing ideas, and see themselves as having a common base of knowledge. For example, curriculum decisions are often made informally, but informal information is shared. Teachers know or find out about new ideas and then try them in the classroom and they keep the school's Curriculum Committee and Advisory Committee informed. Sometimes the Curriculum Committee will poll teachers and get input about new efforts. Internally, teachers emphasize that the norm of reflective dialogue leads to high levels of participation and learning from colleagues:

> It's reflective, it's significant, we bounce a lot of stuff off of each other, you know. It gives me courage to try stuff that I'm cautious to try. It makes me curious and whets my appetite about hey, if they're doing it, then I should be doing it.

Teachers, like the principal, are also used to sharing articles—a clear indication of peer pressure to read:

> I read any article I can get. All of us do and we'll pass those around. Somebody knows I'm having a child that's having a problem. People will give you articles to read to try to help. That's one way we address it.

An experienced teacher, accustomed to using her own reflection as a key to improved practice, remarks on how different her experience is at Agassiz:

> For a long time, the strongest professional development was a dialogue between myself and the literature, and refining it as I shared it with others . . . That's changed totally.

She describes a K–2 staff meeting, where two teachers were new and the others were less experienced than she:

> One of the teachers there presented a lesson plan—and we've been fighting all year long to come up with a reading lesson plan that would reflect Reading Recovery but would be a group situation. She presented one, and Friday afternoon we stayed late, and I said, let's take five minutes and reflect on this. And they took off; it wasn't me anymore, it wasn't me imparting information. And I got just as much from it as they did . . . As we played around with different ideas, I felt I was indeed just as much a learner as anything else.

Perhaps as significant as the openness to learning from each other is the shift in Agassiz teachers' collective sense of themselves as producers of knowledge. This is particularly apparent in their "Restructuring Roundup" program, which provides a weekend conference environment for other teachers:

> All the teachers perform mini sessions . . . on whatever topic they want to do. We have reading, math, and science. Teachers come in from about four or five states. They pay $40 a day to attend sessions. They go to three or four a day. Last year we raised $21,000. We turned away $10,000 because we didn't have room for them. That was our decision to do that. No one makes you do it. If you prefer not to do a mini session, then you usually help. I've done three or four. Every single teacher has done one.

In fact, Restructuring Roundup, according to the Assistant Principal, has become a major focus for the professional development of individual teachers in Agassiz, turning the school into a producer rather than a consumer of knowledge. In the past two years there were not as many in-services for the Agassiz staff, "because Mrs. Cole shared with the group that we were putting so much work into Restructuring Roundup and . . . our faculty was pretty stable with very few new people, [so] she decided we didn't need as many of the in-services." Teachers themselves recognize the magnitude of their accomplishments in terms of how the teacher role is played out:

> Through Restructuring Roundup teachers are building a relationship among each other across the district, state, and region. There is also sharing of ideas among the teachers.

While the required preparation is significant, usually involving most preparation periods and weekends during the spring, the sense of reward for individual Agassiz teachers exceeds the effort:

> It was giving me confidence about what I've done . . . validation . . .
> it was personally rewarding [because] others were interested in what
> we had done . . . you get pumped up for it . . . the experience makes
> you more open to what others are doing . . . it also builds your con-
> fidence and makes you feel enthusiastic.

While Restructuring Roundup is the clearest example of how teachers view
themselves as educators, there are others as well. For example, a growing number of teachers have expanded their role to include teaching lower-income parents how to teach their own children:

> When Agassiz had parent workshops at the school they would not
> get these parents. So they decided to go into the communities that are
> about twenty miles away. The teachers decided to do a parent work-
> shop on a Saturday morning . . . Parents were encouraged to bring a
> child to help demonstrate the learning activities parents can do with
> their children.

For the teachers who have become involved, this effort demands an expanded
repertoire of skills, which they have largely acquired from a dedicated colleague.

What makes it work?
The intellectual vibrance and collective commitment to learning and improving
at Agassiz are not an accident. Agassiz, for example, has a dense network of
informal collaboration that encourages sharing and collaboration between pairs
of teachers. Although these interactions are frequent in Agassiz, most teachers,
when asked, point to more structured settings that foster a sense of collegial
stimulation. Particularly important are weekly grade-level meetings and "faculty study," a monthly meeting that brings K–3 and 4–6 teachers together to think
more broadly about school needs.

> At the beginning of the year, they actually structured time so the
> grade level can meet once a week. It has served the third grade this
> year to do activities across third grade level. And it's been wonder-
> ful because we've integrated the gifted classrooms with the regular
> classrooms. We're actually mixing up the kids . . . The best example
> this year was Famous Americans. Every single child in the third
> grade at this school, doesn't matter what kind of kid you are, is
> studying a famous American. And what we did, we let the kids get
> together by similarities of their people. So if there are five Martin
> Luther Kings, we let them study together.

> Another good thing we do for scheduling teaching and planning, we
> work on grade levels. We meet once a week, sometimes twice a
> week. And we plan the units that we're gonna do together. For
> instance . . . I have gifted children, so therefore I might take mine a
> little bit further than what everybody else is doing. But we all bring

in our ideas. And then we start our planning from there. So it's not just one teacher coming in and having to do it by herself. And I think that's what helps us to grow. It's from the feedback from other teachers.

The grade level meetings are useful settings in which to discuss the specifics of on-going curriculum, but the recently formed Faculty Study Committees have a more explicitly reform-minded intent. More specifically, each of the two committees (K–3 and 4–6) was asked to develop a plan for professional development that was reflective of the concerns of the group. The motivation for the split between lower and upper elementary was that the previous year's discussions revealed that concerns and issues were quite distinctive between the two, in part because the K–3 group is under less pressure to conform to state and district curriculum and testing requirements. The primary function of both groups, however, is to provide safe settings for deeper reflection, with application to the here-and-now:

> It's something different for us because basically we're taking one topic and we're carrying it through . . . I guess we're reflecting on where we are right now; you know, what kinds of things do we find important to us as a teacher, what kinds of things do we want our children to leave us with, what kinds of things do we want our kids to have when they come to us. So I think it's been really important, productive, more so than any of the other things because it applies so much to me here and now.

Equally important is the emergence of informal interdependent teaching roles, all of which are voluntary. Peer observation is one of these. Every teacher is supposed to observe another teacher—on any grade level—for thirty minutes a month. Teachers agree that this is useful, as are the observations that the principal and assistant principal do of them, because they stimulate discussion with others that they do not always know well. In addition to peer observation, which was instigated by the principal, teachers have, on their own initiative, developed other ways of sharing information and collaborating in the classroom. Both of the special education teachers, for example, work with some teachers collaboratively in the classroom, an approach that they initiated after attending a conference on inclusion several years ago. A "Teach for America" faculty member team-taught with one of the school's master teachers for her first year. Several teachers identified themselves as classroom "mentors" or were identified as such by others.

Finally, one cannot ignore the pervasive effects of the school's culture, which were carefully cultivated by administrators. Although the district was strapped for funds, virtually all Agassiz teachers say that they spend their own money to go to conferences and workshops that can hone their skills. As the principal explained, all of the activity in the school has "made us stronger, ravenous for knowledge . . . to have knowledge before we start changing things."

Organizational learning in the Okanagon school

Okanagon in context

In 1981, the Okanagon Community School, a 6th–8th grade middle school located in a poor residential section of one of the largest metropolitan school districts in the U.S., closed its doors after rumors of its ineffectiveness caused its enrollment to plummet from nearly 3,000 to 600. To the community's consternation, "their school" was replaced by a magnet school focused on the performing arts, which attracted few local students. In 1990, the Okanagon Center for Advanced Academic Studies reopened as a non-selective "neighborhood magnet" school, after a year of planning by a team consisting of several teachers, the principal, and representatives of social service agencies, citizen groups and local business. Driven by a dream of educational equity and opportunity for poor urban children, the school enrolls nearly 1,500 students, including approximately 37% African-American, 33% Filipino, 20% Hispanic, 5% recent Southeast Asian immigrant, and 8% white students. Nearly two-thirds qualify for Chapter I support, and over half receive free lunch.

"The dream"

Okanagon was founded on the assumption that dedicated professionals could outwit a system that had failed to create a "level playing field." According to Bill Stone, the principal, the school's vision is both educational and political:

> One advanced curriculum for all children in three years, so they can
> go to any high school in America and not get tracked out of the
> power class.

A significant part of the curriculum is centered on interdisciplinary units. The teachers collaborate on the units and work to fit the curriculum to their students' learning needs. In addition to the focus on content relevance, the school has established "The Okanagon Standard," which consists of required student performance "challenges" that range from community service to the completion of a research project. Although the school spends considerable time on curriculum development, the faculty remains attuned to the state's curriculum standards and assessment procedures:

> We try to do our best on the statewide assessments. We try to honor
> those things but I think that [other measures of performance are]
> more important. What the kids come to school with when they get
> here and what they do while they are here. And also what they do
> once they get into high school.

In addition to the emphasis on academics, teachers and administrator argue that their students need strong social and emotional support—in part because of their age, and in part because many of them come from family and community contexts that do not provide enough security and certainty. To "make big, small" the school is divided into nine "Families" that form the basis of both

teacher and student life, and are the primary unit for day-to-day work and decision-making. All core academic teachers (social studies, math, language arts and
science) are assigned to a Family; the remaining specialist teachers (special education, band, languages, etc.) are part of the "Discovery" Family:

> The people closest in proximity to teaching and learning must have
> most of the power . . . Basically, we say there are two levels of deci-
> sion-making at this school: the educational Family and the
> Community Council. And it's sacred, that all decisions that have to
> do with teaching and learning are made only by Families.

During the planning year, the teachers decided that they would take on roles
normally carried out by administrators and specialists in return for more teachers and smaller class size. In practice, this means that each Family has discretion
over staffing, schedules, use of paraprofessionals and special education
resources, and, to some extent, curriculum; the Family Leaders play a central
role in the life of the school:

> I am a Family Leader, and as Family Leader, I am the representative
> on the [Community] Council . . . I am a mentor teacher, and I work
> with new teachers, student teachers, teaching assistants . . . every-
> body talks to me, but I [also] talk to everybody. It's a key role in that
> I can keep a handle on the heartbeat of the school and the morale of
> what's going on . . .

It is not just Family Leaders who see a difference in their roles at Okanagon. The
expanded roles also have significance in working with students, creating closer
ties with students that reinforce "The Dream" that the school will have a fam-
ily-like character:

> At our school we [teachers] are also counselors and administrators .
> . . My complaint early on in my teaching was that a math teacher
> didn't have an 'in' [because students didn't write about themselves in
> my classes] . . . At this school, since I'm a counselor, I get to see inside
> the children and . . . I understand them more.

Teacher intellectual leadership and teacher learning
An unspoken aspect of The Dream is that faculty must take responsibility for
the quality of professional work in the school. According to most, this is a never-
ending task that requires that all faculty read, attend conferences, reflect, and
share ideas. The faculty uniformly point to the ways in which the school is connected to ideas and reform networks outside the school that provide stimulation
for their own evolution:

> I really believe that [Okanagon has helped me to plug into the major
> educational issues of the times]. We are part of the Coalition of
> Essential Schools, Brown, the New Standards Project, PACE, my

portfolio group here [in the district]—these are the ones that I am directly involved in myself.

Teachers point out that the opportunities for involvement are so numerous that they must place limits on them or risk cutting into important time with colleagues and even time to develop the curriculum at the school. The school has also developed norms that ensure that most teachers are involved in some external activity, and that no teacher becomes a "star:"

> Those corporations that we're involved with will say to us, "Oh, send Barb again," and we will say to them "No . . . the Okanagon Way is to allow everybody to have an opportunity [to go to national meetings]" because we believe that when we talk about what's important to us then it helps us remember what's important.

And, they are careful not to make the national trips a "perk," but to view them as a responsibility to colleagues:

> What we have done is [use time in faculty meetings] as another means of being accountable for those of us who go places. There is a responsibility to give [information] back, which helps us to professionally grow too.

As this implies, teachers value the important connections with groups outside of the school, but they also learn a great deal from each other through planned and sustained interactions. The staff view each other as expert resources, and often provide each other with both formal and informal professional development opportunities:

> I've been doing the training in Socratic seminars since 1989. It's incredible to work with [other] teachers. I love that. People can have new ideas, pick your brains, and you learn back from their giving you a scenario and it makes me think . . . and in that way I feel like I grow too because I can't say, "Oh I have all the answers." (t1:ok)

Okanagon also has ongoing self-study activities in a number of school-wide committees. For example, there is an Evaluation Committee, whose responsibility is to collect longitudinal data about the school's performance. The school's demands for information are often an embarrassment to the district, which finds that, in spite of a large research staff, it cannot meet the needs of the data-hungry Okanagon staff. In response, the school has begun to initiate its own research and assessment, and to develop its own standards of performance:

> The other method of assessment that we are working on is [our own] school-wide assessment, where everyone writes on the same topic for language arts and everyone receives the same math problem. The whole staff sat down to determine what a 1 looked like and a 2 and a 3 and a 4.

The teachers developed these standards for assessing performance in interdisciplinary groups—it was considered an important development for teachers to understand and contribute to what constituted a high or low level of performance in mathematics, even if the individual was not a math teacher:

> We are counselors to these students, and we keep the results of those tests for our students . . . By doing this, I think we set an example to the kids that all teachers are of one accord (about performance standards).

After the students were tested, teachers graded the papers of students from other teams, and the results were shared with the entire school. Teachers seemed unconcerned about this public insight into the performance of their own students:

> Well, it could [cause concern], but I think that we all look at it like a training thing for us. I mean, if for some reason all of your kids didn't do too well, then we can help each other.

> I think that the level of comfort about exposing dirty laundry and just talking honestly about what is going on—the key is honesty, not just "show and tell."

> The [school-wide assessment] is a pain in the butt but I like it. It gives you a sense of where the kids are, relative to the whole school. It gets the whole school involved where the strengths and weaknesses are.

Time for more informal discussion is allocated to the Families. Teachers schedule time during the school day for Family meetings. The meetings are informal times in which teachers can engage in reflective dialogue and discuss what the curriculum looks like to them. Even within school-wide projects each Family is given the decision-making power to model the project to fit their students' learning styles. Time to meet and talk, as well as an emphasis on communication, enables the members of Okanagon to enjoy a sense of support and community. For example, in one meeting the Dragon Family teachers talked about the criteria for judging the research papers that the students are required to write to earn the Okanagon Standard. The meeting didn't resolve the issue, but they ended by deciding which team members were going to develop rubrics to score the papers. Not all meetings focus on educational issues, but most teachers pointed to the value of the time:

> Our Family spends some . . . planning time dealing with children and discipline, and some amount of time scheduling parent conferences, but I would say 75% of our time is spent on the kinds of issues that we set up this time for.

The principal's comments about the teachers summarize the conditions that most believe they work with:

There's a passion here for teaching . . . Different people are emerging as leaders, and . . . are taking their gifts and making them go, and other people are learning from them, and there's collegiality and co-teaching and peer evaluation.

Making Okanagon work: The challenge of autonomy and coordination

Okanagon is a large school, and the initially simple structure—Families and a Community Council—worked well only during the initial year of operation, when the school had one third of its full enrollment, and most of the teachers had been involved in the initial planning process. Now, teachers and other key observers agree that more limited coordination and communication are barriers to achieving The Dream.[4] Some of the issues are temporary, and related to the school's rapid growth, such as the problem of socializing new teachers to The Dream. Others, however, are more permanent. One of the key features of the Okanagon Way is that Families are responsible for making decisions about their own organization and functioning. While most of the Families appear to function effectively, the accumulation of independent decisions has fragmented teachers' understanding of the school's mission:

> I think that the school as a whole is trying to bring children to high academic levels. I am not sure that we all agree on what that is. And I am not sure that we agree on how to do that.

The need to develop a new way of approaching the education of poor, inner-city children, where few successful models are available, also strains teacher resources and increases differences between Families:

> So the curriculum is kind of more winging it in picking and pulling pieces and putting them together, which is really exciting on the one hand but it's all on my shoulders or my team mates . . . There are some good things being developed, but none of it really, really matches what we see our needs as.

A "friend" of the school in the district office commented succinctly that:

> One of the big issues at Okanagon is communication, and making sure everybody has common understandings and access to information, and even the opportunities to be together. The curriculum is being delivered; all kids are getting the same curriculum, it is advanced. What I would wish for Okanagon is that there would be a way to make sure that teachers could have more time to meet and make sure they really are on track with each other.

---

4 Okanagon faculty hire new teachers for the school. Although the principal serves on the hiring committee, he has only one vote. They spend a great deal of time in the hiring process explaining the "Okanagon Way" to make sure that prospective teachers are willing to commit to their "Dream" and the extra effort it involves.

Teachers noted that the informal system of sharing between Families was bolstered during the first three years by foundation-funded retreats, where emerging Family issues were discussed, and some areas of consensus were found. More recently, the school's committee structure has begun to serve a more formal coordinating function.

The Curriculum Committee, which consists of members from each of the Families and the principal, was initially intended to propose school-wide projects. More recently, however, it has taken on an important policy-making role beyond its original mandate. For example, the decision to engage in school-wide assessment activities was initiated by them. The Committee and its subcommittees (such as the Portfolio Assessment Committee) are widely viewed among the staff as more influential than the Community Council, in large measure because they meet more often and discuss more substantive educational issues. Family representatives to these committees report back to their unit and the Families are then expected to collaborate on the school-wide projects. Teachers agreed that this line of communication is essential to school coordination, but there is a lingering sense that the lack of direct participation in making important school-wide decisions may be undermining a sense of common purpose. In particular, the Curriculum Committee, because its work is so time-consuming, has attracted the teachers who are most intensely concerned about improving school-wide academic performance. Some teachers, who view the Family structure as more consistent with the socio-emotional support for teachers and students, regard the Committee's initiatives with some concern:

> And the [school-wide] assessments have been in addition to everything that we are doing in the families toward the [other school-wide] projects. So what we meant it to be and what it actually came out to be are two different things. So we got a lot of people mad because we have "one more thing to do."

> Well [there's been] some [resistance]. But part of it is that you forget why you are doing it [because of all of the other things you're doing].

The growing role of school-wide coordination of curriculum and assessment is not the only conflict between school-wide versus within-Family concerns. Okanagon faculty have intense debates over where The Dream should go: Among the issues are whether there should be cross-age grouping, whether students who are severely learning disabled should be fully mainstreamed, and whether Families should have complete freedom to determine their schedules, even when these may create problems for the cross-Family "Discovery Team." Other teachers express concerns that the business of restructuring, even though the teachers are all striving toward The Dream, may distract from the central task of teaching children who have many needs and few resources outside of school:

> Well, I never had to be the guidance counselor. I can say that today there were four instances where I was pulled out of my class for different things.

> [Because teachers do the administrative work] it's sometimes diffi-
> cult to run around and get information, because it's spread over the
> whole school. Sometimes one job depends on another person's job,
> so it becomes a monumental task just to get everybody together . . .
> Even things like arranging for me to attend a conference might
> involve me seeing three people as opposed to one . . .

Other teachers worry that the school has yet to demonstrate that it can raise the
performance of the school's disadvantaged children on any of the state's external assessments, and although they believe that their students are doing better
in high school there is still feedback that they are not fully prepared for the higher levels of academic work.

Teachers at Okanagon debate these issues fiercely, and in open forums,
including retreats, "Open Forums," and school-wide committees. Teachers
agree that the school belongs to them, and that all of the constraints on their
freedom are those that they collectively choose. Nevertheless, there is ambiva-
lence about whether the ongoing learning experiences are a stimulus (as most
agree) or a constraint (as most also agree):

> I think that we do have a lot more freedom to do things in our class-
> room [when compared to a traditional school] but it is harder some-
> times [for people not involved in making the decision] . . . Because
> my Family teachers weren't there, [when a decision is made] there is
> a problem when it hits them for the first time. It sounds and feels a
> lot like what happens in our old schools like "You have to do it like
> this." You know what people's reaction is to that!

But, as faculty noted, at Okanagon there is no certainty except that things will
be changing for the foreseeable future. Teachers who are not willing to live with
that may leave (as several have done); those that stay express a certain resigna-
tion about the ever-unfolding nature of the school:

> I just have to trust. I have got to trust that they (members of other
> committees) are about the same issue as I am—educating children—
> and that maybe this is the best . . . and then the opportunity is always
> given to say, "What do you still have a problem with?" and those
> issues are put up also. And maybe someone in another group will say
> "this is how I see it." And if there is no resolution to those issues a
> subcommittee is assigned to see if we can resolve those issues . . .

Learning about organizational learning from Agassiz and Okanagon

How can we interpret the apparent success of Agassiz and Okanagon in developing learning communities that are focused on student achievement? In the
remainder of this chapter we will compare and contrast their accomplishments
using elements of an organizational learning framework, and go on to elaborate

some additional features of their structure and human resources that help to explain how they were able to develop learning capacities.

A model of schools as learning organizations assumes that adult development often takes place in groups and cannot be reduced to the random or systematic accumulation of individual knowledge (Louis, 1994). In a learning organization, the technical knowledge base—inclusive of content information, teaching methodologies, and innovations in both areas—provides structure and context for creating new social and intellectual bonds. Teachers focus on their own individual learning about both teaching and content areas but also discuss, share, and critique so that all members understand and can use the newly learned knowledge. Thus, when we think of school change initiatives, the organizational learning perspective suggests that the collective, regular processes of teachers and administrators working together around issues of practice and professional knowledge will provide schools with the capacity for change and development.

In explicating our ideas about organizational learning in Okanagon and Agassiz, we will focus on three key concepts that have not been well developed in more commonly cited literature (Fullan, 1994; Senge, 1990):

- organizational memory
- the knowledge base for learning
- information distribution and interpretation

Our discussion is based on the premise that individually held information or knowledge is of little use to the school unless it is retrievable and understood by others, and that common knowledge must extend beyond individuals and specific points in time to be of use to subsequent "generations" of members of the learning community. It is this process of storage and retrieval of information, and the collective feeling which accompanies learning, that lie at the heart of organizational memories that contribute to furthering effectiveness.

Organizational memory

We regard both Agassiz and Okanagon as schools that have, in many ways, demonstrated conformity to the definition of a learning organization derived from theory and research (most of which is based in the study of private sector organizations.) However, there are also some differences between the image of organizational learning presented by these schools, and what one would expect based on theory alone.

Shared memory

This consists of collective understandings that are developed in an organization over time. The shared memories held within a school will influence its capacity to learn (Louis & Miles, 1990). Positive shared memories from previous learning situations create an openness to future learning; conversely, memories based on bad experiences act as barriers to new learning efforts. Without an adequate base of common understanding from which to draw, teachers can be reticent to begin new learning activities (Louis, 1994). Turnbull (1994) points out that the postmodernist assumption that all knowledge is local refers to its production, and not to its distribution and/or use by others. He illustrates the importance of memory in his discussions of the various mechanisms—artifactual and oral—by which pre-literate societies were able to achieve complex scientific feats, such as the building of cathedrals over several centuries.

Unlearning as a key process
Considerable attention is paid in the research literature to the importance of collective memory. It is viewed as critical to the ability of the organizational members to draw upon it both to define issues that need to be addressed and to assess promising solutions based on previous experience. These cases, however, show only a limited importance associated with recollections of a common past. In the case of Okanagon, the fact that it is a relatively new school may account for this: there is only a little history to drawn on. But lack of interest in or reference to the past also characterizes Agassiz, which is a school with a continuous history in a district whose history is both dramatic and contains important discontinuities associated with desegregation and changes in the district office and state.

When the interviews are examined, a reason for the lack of significant attention to memory emerges. The teachers in these schools are preoccupied with "unlearning" the past, rather than drawing on it. Most, when referring to the past, cite experiences in other schools or at other times that they believe exemplify poor practice, structure or values. What they point to is the fact that they are on the cutting edge: they create new models through constantly re-examining their own behavior for pernicious evidence of "traditional thinking" rather than looking to previous experience as a guide.

In education, this may be a condition that most learning schools will confront in the next decade or so: learning must occur, but there are few images that teachers can draw on in either their own or their school's experiences that will help them to move forward. Where "unlearning" is more important than memory, the process of using the past to guide the future is significantly different.

"Organizational newness" as a stimulus to short-term memory
Although neither of the schools paid a significant attention to the use of collective memory, more teachers and administrators in Okanagon, which was a "new school," referred to its "history" than at Agassiz, which was a school where change had evolved somewhat more gradually. In particular, many people at Agassiz referred to the planning year (in which six of the teachers and the principal were key actors), The Dream that stimulated the school's founding, and some of the key events in the school's short history, such as the problems of hiring and socializing many new teachers in a short period of time.

One explanation for this surprising finding is that new schools may be more dependent on past reference points that exemplify "what we are about" precisely because the teachers must develop common understandings in a very short period of time. The case studies do not suggest that Okanagon was more cohesive than Agassiz—on the contrary, despite all of the references to The Dream, teachers were open about the fact that reaching a satisfactory consensus in the school was an arduous and uncompleted task. Having a common past and a dream served as a mechanism not for problem solving, but for building "we-ness," in sentiment, if not actual agreement.

The knowledge base: using internal and external sources of knowledge

Learning has typically been defined by the notions of acquisition, storage and retrieval (Lave, 1984). Organization learning expands on this definition in that knowledge is considered to have a shared, social construction for many members of the school community (Louis, 1994). Organizational learning and individual learning are clearly interrelated, but while the latter may occur in the absence of social interaction, the former is dependent not only on interaction, but also on the affirmation of common learning goals. A school's knowledge base is drawn from three basic provinces within the organization—individually held knowledge, knowledge gained through school-wide self-appraisal, and knowledge that is available from outside the school.

Individually held knowledge

Teachers enter schools equipped with broad, often disparate foundations of information related both to content and pedagogy. In many cases, however, individually held knowledge is difficult for colleagues to access and utilize, even when efforts are made to create arenas for sharing (Hargreaves, 1994; Little, 1990). Further, the endemic uncertainty of teaching exacerbates the problem when teachers work in settings that are characterized by limited levels of trust and communication about professional issues—even very knowledgeable teachers may not believe that they have valuable ideas to share (Louis & Kruse, 1995). In traditionally organized schools, for example, teachers do not have access to forums that encourage sharing of expertise (Lortie, 1975). Staff development is usually provided by outside experts, while staff meetings are brief and dedicated to administrative issues. Even in schools where staff development funds are generous, but focused on individual learning, teachers may be unaware of the efforts that others have made to develop special knowledge in content or instructional methods.

In both Agassiz and Okanagon, on the other hand, explicit attention is paid to honoring individual knowledge. Teachers are not jealous of each other's expertise because they do not assume that they are all equal in all ways. Rather, they talk about each other's special skills or knowledge with admiration—as additional evidence that this is a school that has a strong professional character. Teachers who are acknowledged by others for their specific expertise become resources to their colleagues. When teachers at Okanagon participate in individual opportunities, they are obligated to train their colleagues in the areas where they have benefited—not just report back on their experiences in a full-agenda staff meeting. In Agassiz, the "Restructuring Roundup" is a celebration of individual teachers' particular competence—and teachers learn from each other as they share in the preparation for the sessions. Individual development is central to their processes of organizational learning, rather than a byproduct or an ancillary activity.

Knowledge created through self-appraisal

Using a constructivist perspective, knowledge is created when the school self-conciously examines its own work. Teachers at Okanagon and Agassiz work together to consider what they are accomplishing in the current curriculum and instructional practices, which stimulates new frames of reference for performance. While self-appraisal efforts are often generated by pressure from district or state accountability initiatives, school faculties may also use these external stimuli to focus on internal assessment. In Okanagon, for example, the state's accountability and testing initiative are important to the school. The internal assessment of student performance using their own test was stimulated by this, but went far beyond. Discussions of the commonly-held knowledge generated by faculty discussions about the purposes and goals of schooling and student performance on the internal assessment provide a common vocabulary which facilitates further collaborative learning.

The presence of structures and routines that support learning are important to self-appraisal endeavors, and provide a framework for acquiring information about what new ideas may be important to consider (Cohen, 1991). In Agassiz, for example, the Faculty Study Committees were not only a means of "keeping up," but also of reflection and dialogue around specific issues of practice in their school.

Knowledge gained by organized search efforts

Finally, we suggest that schools can pursue deliberate efforts to gain information pertinent to reform. As schools collectively search for solutions to identified problems, knowledge is generated that both further defines the problems and suggests alternative actions. Search efforts depend on the capacity of the organization to absorb and use new ideas to create alternative organizational structures and ideas (Louis, 1994).

Both of these schools are creating a knowledge base that is fed by both internal and external sources of knowledge. In both cases, individual teachers, and small groups, are actively involved in external networks, ranging from local groups (such as Agassiz's work with other schools in the district on Reading Recovery) to multiple national groups (such as the Coalition of Essential Schools and a variety of specialized associations). Teachers *read* in these schools: they do not just attend courses. This is most clearly exemplified by the Faculty Study Committee at Agassiz, because it formalizes the teachers' vision of themselves as connecting directly with the ideas of others, but is also evident less formally at Okanagon. Notable in both schools is the assumption that teachers have a responsibility for sharing externally acquired ideas with peers, not superficially, but by demonstrating, running seminars, and other ways of increasing teacher-to-teacher communication. Okanagon, in particular, is deeply committed to getting information about its performance: from parents, from the high schools that its students attend, and even from standardized tests (most teachers concur that the new state tests are closer to the kind of authentic learning that they want their students to demonstrate than previous tests).

Improvement initiatives based on one information source are rare. In fact, the case studies suggest that relying on only one of the three knowledge bases can be detrimental to the overall health and longevity of the school culture. Teachers are reliant on ideas from outside of the school, but also see themselves as a collection of experts whose knowledge is available and can be drawn on. Sharing expertise is an important component of both schools' ethos. Furthermore, school organization and construction of a commonly held knowledge base is neither a singular activity, nor one that is ever completed. Knowledge base development is, rather, tightly-linked to information distribution and interpretation.

Learning processes: Information distribution and interpretation

Knowledge from different sources is shared and interpreted when it is matched with existing organization frameworks or schema (Huber, 1991). In schools there are multiple sources for interpretation, such as belief structures, frames of references and organizationally based social constructions, the cues and feedback related to organizational symbols and shared understandings. In Okanagon and Agassiz, information distribution and interpretation can be best thought of as focusing on the use of information to enhance shared understandings.

Information distribution is not the placement of photocopied articles in teacher mailboxes, nor is it the transfer of unquestioned information from one teacher to another. Instead, information distribution and interpretation is the construction of meaningful contexts and conditions under which new routines are practiced. Individual effort is required to learn a new curriculum or instructional methodology, but learners are acquiring not only explicit "expert knowledge" of their own, but also the embodied ability to behave as members of the school community. Where schools move to more communal forms of practice it is important for teachers to know and understand the instructional and curricular foci that are common to all teachers. It is by knowing and understanding the foundations of the new innovation that teachers can subsequently translate the new information into useful classroom knowledge and practice that is also shared across classrooms.

Assumption sharing and interpretation

Teachers at Okanagon and Agassiz rarely assume that when an issue comes up they will automatically know what other teachers believe and are doing. Rather, they view the development of a common set of assumptions as problematic, and needing constant reinforcement. A significant example from Okanagon is the concern expressed over the need to maintain common assumptions about the meaning of the "Okanagon Standard" (required demonstrations of competence that students must carry out before they graduate) between Families, while at

Agassiz the continuing discussions about the role and importance of standardized testing demonstrate the teachers' commitment to continuous collective wrestling with difficult issues.

But teachers in both schools do more than share assumptions, which often occurs in more open-ended discussions: they also spend time on the more mundane tasks of making sure that they agree about specific information that they have about their school. In Okanagon, for example, teachers had to work very hard to develop a consensus about how to score the school-wide assessments that they developed, and spent a lot of time discussing the meaning of the results. In Agassiz teachers regularly brainstorm about indicators of success at each grade level, as well as for the school as a whole, and are struggling, slowly, to interpret and deal with the troubling absence of African-American children in the state-mandated gifted program.

Systematic versus incremental learning
The research literature from the private sector suggests that all organizations learn incrementally. What distinguishes organizations that only learn slowly and incrementally (Senge's "boiled frog" analogy) from a more dynamic learning environment is the emphasis on more formal efforts to learn. In both of these schools, there are structures that teachers have developed themselves to encourage systematic learning. At Agassiz, the grade-level meetings focus on specific learning related to the curriculum, while the Faculty Study Committees are struggling toward a common learning process. The Restructuring Roundup, although it emphasizes the role of individual teachers in creating workshops, also presents an important opportunity to reinforce pulling together "craft knowledge" in ways that make it more visible and accessible to colleagues. At Okanagon, the emphasis on systematic learning is more clearly structured, with many powerful committees—the Curriculum Committee, the Evaluation Committee and the Portfolio Committee—approaching different aspects of the need to engage in self-examination and decision-making.

Paradigm shifts
In both of these schools there is evidence that teachers think about their work very differently from typical teachers. In both schools, for example, most teachers think about themselves as peer-teachers as well as teachers of children. In Agassiz, the more extensive role of the school in parent education is also emerging, while at Okanagon the teachers' recent efforts to design and score school-wide assessments reflect a sea-change in their sense of collective responsibility for student learning. The recognition of these shifts increases teachers' sense of excitement about learning: they see themselves as challenging what others think and believe about teachers and inner-city schools.

Some additional differences between the two schools

Although both Agassiz and Okanagon share important characteristics of organizational learning, the brief case studies also reveal differences between them
that suggest that there will be many models for what "learning school" may
look like.

Leadership styles: Internal mentoring versus external sponsorship
The leadership in both schools was viewed as supportive by teachers, and instrumental in achieving the educational goals of the school. In both schools, leaders were powerful actors in the intellectual development of the school. The
enactment of the school principal role was strikingly different, however. In
Agassiz, the principal (and the assistant principal to a slightly lesser extent) were
viewed by teachers, and viewed themselves, as the source of ongoing intellectual leadership on a daily basis. Teachers are grateful that leadership retains a constant focus on the identification and resolution of educational dilemmas; they
refer to the principal often in their interviews as the source of ideas to resolve
issues that are perplexing and difficult. In addition, they view their own empowerment as contingent: it is still "granted" by a benevolent and wise individual
rather than something to which they are entitled.

In Okanagon, on the other hand, the power of the principal's intellectual
vision is also acknowledged by all teachers, but he was viewed as a touchstone
rather than a problem-solver. Ideas that were too far "out of line" with The
Dream (for example, when a few teachers wanted to have a substantially separate program for severely learning disabled children) stimulated conditional
warnings that a particular line of thinking was inappropriate. Yet, although he
was the "Keeper of The Dream" (a title that appeared on his name tag in lieu
of his formal position), most teachers viewed him as a distant influence on day-
to-day work in the school. They saw his role, most importantly, as a publicist
for the school, and as an astute politician who could keep their efforts at once
visible and protected. When asked what would happen if he left, most teachers
felt that they could find another leader from within their own group who could
carry on with the job; they did not see their empowerment as contingent on him
as an individual.

Size and its consequences
Although Agassiz is not small for an elementary school, it is less than half the
size of Okanagon. Both schools have found it necessary to create structures for
communication that are smaller than the whole faculty, but the problems of
communication induced by size and its associated complexity are striking. In
Agassiz, the need to form smaller and more intimate groups is met through
grade-level meetings (approximately six teachers in each of six grades) and the
division into two Faculty Study Groups of approximately twenty teachers each.
It is too early to determine whether the division of the school into two "conversations"—one for the K–2 teachers, and the other for the 3–5 teachers—will

create internal tensions. Yet, there is a strong sense of the whole that is expressed by most teachers.

In Okanagon, a structure that includes ten "Families" and eighty-four teachers is far more complex. (Although Okanagon has subject-matter departments, they are not an important component of professional identification.) Problems of communication, particularly related to issues of sharing assumptions and interpretations, surfaced shortly after the school expanded in its second year. Okanagon's solution was the development of a complex system of committees that have overlapping responsibilities, and which represent Families. However, because the school's smaller committees often make critical decisions, the issue of how to gain ownership and ensure quality communication to others in the school is at the top of many people's minds. In large measure, emerging frustration with the apparently endless committee work among some teachers (typically those who are not on the major committees) is a consequence of the difficulty of creating a unitary democracy in an organization that is too large to function as a whole.

Conflict management
All schools have conflicts, whether they are interpersonal, intellectual, or whether they concern power and its consequences. In both Okanagon and Agassiz, the controversial issues, according to virtually all teachers, revolved around educational ideas, providing strong support for the assertion that these are learning communities. In addition, in both schools there was agreement among teachers about the content of contentious issues: teachers in Agassiz, for example, tended to agree that the issue of low African-American representation in the gifted program was of concern, while, in Okanagon, teachers pointed to the tension between the desirability of school-wide coordination, and the demands that it made on teachers. On the surface Agassiz appears to be more serene, while Okanagon teachers are more forthright about conflicts. We attribute this in part to "norms of politeness" in the deep South as compared to the West, but more importantly to the relative ease of resolving many disputes in Agassiz as contrasted to Okanagon. In Agassiz, all teachers agreed that, in the end, the principal would adjudicate disputes that could not be addressed within the staff; in Okanagon, although some teachers wished that their principal would take on a conflict resolution role, they were forced to create a consensus among themselves because the principal refused to do so unless The Dream was at stake. Again, although many teachers talked about the importance of "trust" and the willingness to expand their personal "zone of indifference" to permit colleagues to make binding decisions, there was a longing for an impartial judge, even among several of the most influential teachers.

Some additional structural similarities between the two schools
While there are clear differences between the two schools in terms of what Organizational Learning looks like, there are also some critical similarities in how the schools are structured, and how human resources are used to support this.

Time, time, time
In both schools the structure of the school day, week and year is organized to make sure that teachers have time to work together. The amount of time spent in meetings is substantial, and varied in terms of length, purpose and focus. Short meetings of teams or grade levels occur several times a week, and often cover logistics (how to teach a unit that they have decided on in common). Longer meetings occur in committees or in specialized professional development context, often after school. Both schools have more extended, all-day professional development experiences where the whole staff engages in activities that focus on learning, especially on examining assumptions and generating consensus about mission and values. In both schools teachers also talk about informal interactions, but these play a small role compared to the formal contexts— including agendas and expectations of action—that have been developed. There is no one solution to the problem of time, except to note that teachers need times of different lengths, devoted to different purposes, to achieve the kind of learning demonstrated in these two schools.

What does "real empowerment" mean?
An outsider's assessment of empowerment would probably conclude that Okanagon teachers had more formal authority in their school than did Agassiz teachers. This conclusion, however, must be viewed in context. Agassiz is located in a district and state notable for their history of a "top down" and bureaucratic approach to management; Okanagon, in contrast, is in a city and state that are more characterized by efforts to firmly direct and support school renewal and the building level than to regulate. In Agassiz's context, teachers felt free and entirely professionalized—as they did in Okanagon. The belief on the part of most teachers that they could make decisions about the curriculum, school organization and staffing that were consistent with their best judgement about the educational needs of the children in their school was palpable in both contexts.

Multiple structures and strategies for creating interdependence
Neither school had formal team teaching arrangements, or requirements that teachers engage in peer coaching or other forms of interdependent teaching behavior. In both schools, teachers believed that they still had room for total individuality in their classrooms—there was no sense of "contrived collegiality." Yet, in each case there were also both opportunities and arrangements that encouraged interdependent teaching, which included observation of other teachers' classrooms, voluntary teaming for short or long periods, and structures that permitted (but did not require) the coordination of teaching strategies and materials. Again, as with the multiple structures for encouraging conversation and reflective dialogue, these opportunities were used by teachers when they suited, and ignored when they did not. In Agassiz, for example, teachers were strongly encouraged to observe each others' classrooms on a monthly basis, and many did so, and cooperative learning could also take place as people prepared for Super Saturday. Grade-level teams occasionally prepared common student expe-

riences, but did not feel pressured to do so all the time. In Okanagon, the Family structure provided easy opportunities for collaboration and "safe" interdependence: several teachers reported that the ease of working with each other in more intimate ways was growing over time. In addition, in Okanagon, special projects were used to create interdependence—for example, the effort to get teachers of all subjects involved in determining how to assess student performance on mathematics and language arts problems. Again, however, interdependence was voluntary and not mandated.

Adding new roles for teachers
In both schools, teachers were assuming significant roles beyond that of classroom teaching. These were quite different in the two schools, ranging from counseling and assessment development in Okanagon to peer professional development in Agassiz. In both schools, however, teachers noted that expanded roles were, by and large, sources of expanded satisfaction with their work, and of pride in their professionalism.

Some additional human resources similarities between the two schools
The above discussion highlights some of the changes in structures, role definitions, and distribution of power and authority that are associated with organizational learning in the schools. In addition, teachers in both schools noted other key features that are more associated with changing the culture of the teaching environment.

Recruiting and socializing teachers
Although one might view the opportunity to interview and hire teachers primarily as a feature of the power relations in schools, teachers talked of it more in terms of opportunities to ensure that their colleagues would share the same values, and be willing to work within a framework that they themselves had designed. At both schools, teachers talked about the recruitment process as an opportunity to present the school, to explain the philosophy, and to be clear about peer expectations. The goal in both settings (where schools were obligated to take teachers whose jobs had been eliminated in other schools) was to ensure that those who would not fit in would choose to go elsewhere. For the most part they believed that they were successful, and that they provided the kind of mentorship and support to new teachers that would help them fit in (and, for the most part, newer teachers agreed that they had been supported). On the other hand, teachers in both schools were clear that when teachers did not fit in, or were not willing to work to the expectations of the school, they would not necessarily make it easy for them to "opt out" or get by. Both schools reported that teachers who "didn't fit in" left voluntarily.

Peer pressure
As implied above, both schools exhibited strong indications of peer pressure. When teachers were asked who they felt accountable to, a number reported that

they worked to the expectations of their peers—to meet the norms of the group, and not to let people down. A new teacher at Agassiz, for example, reported that the school had a reputation of "weeding out" people who weren't willing to work hard. Another teacher, when asked what teachers with young children did about the pressure to work long hours, reported that they weren't expected to stay after school, but they would, instead, come in on the weekends!

Effort and its consequences

Much has been made of the problem of "burnout" among teachers in restructuring schools: too much work, too much pressure, and too little support. No teachers in these schools reported being burned out. Several indicated that they had, in the past, felt overextended, and had responded by getting off of one committee, or resigning as head of a task force. Other teachers were in line to take their place as they achieved a different balance. Yes, teachers worked very hard in these schools; far more than the contract, and far more than many professionals in other fields. While we have no hard data, it seems clear that the average teacher that was interviewed in this study was working more than 50 hours per week, and often put in extra time in the summer (the latter was reimbursed in many cases). Yet, despite the fact that many of them had been putting in the kind of extra effort necessary to create a learning organization environment, most seemed energized rather than tired. We believe that this is because the additional effort was voluntary, intellectually stimulating, and associative—that is, connected with other people. Under these conditions, the amount of work was rarely considered, except in passing: it was part of the enjoyment of life. The old adage about dying men never saying that they wished they had spent more time in the office would not apply to this group: most would say that the time that they spent at work was both professional time and time in service of a higher goal.

Reliance on external networks for professional development

These two "learning schools" are clearly open systems: they rely not only on infusions of additional funds for retreats, computers, professional development opportunities, etc., but for a sense of connection with a higher agenda: the reform of U.S. education. Even when teachers indicated that they believed they had more to contribute than they obtained (for example, one Okanagon teacher remarked with amazement that he and close colleagues had examined portfolios from many different schools and had found that most lacked any criteria for what constituted an effective portfolio task), they still found that working with other like-minded people from throughout the country on portfolio development was important. The learning organization is critical to teachers' professional development and school change, but schools need intimate connections with a broader agenda to reinforce the energy needed for their hard work.

References

Berends, M. & King, M.B. (1994). A description of restructuring in nationally nominated schools: Legacy of the iron cage? *Educational Policy* 8 (1): 28–50.

Cohen, M.D. (1991). Individual learning and organizational routine: Emerging connections. *Organization Science* 2 (1): 135–139.

Daft, R. & Huber, G. (1987). How organizations learn. In N. DiTomaso & S. Bacharach (eds.), *Research in Sociology of Organizations*. Vol. 5. Greenwich: JAI.

Fullan, M. (1994). *Change Forces*. New York: Teachers College Press.

Hargreaves, A. (1994). *Changing teachers, changing times: Teachers work and culture in the postmodern age*. New York: Teachers College Press.

Hedberg, B. (1981). How organizations learn and unlearn. In P.C. Nystrom & W.H. Starbuck (eds.), *The Handbook of Organizational Design*. Vol. 1. New York: Oxford University Press.

Huber, G.P. (1991). Organizational learning: The contributing processes and the literatures. *Organization Science* 2 (1): pp. 88–115.

Lave, J. (1984). *Everyday cognition: Its development in social context*. Cambridge: Harvard University Press.

Little, J.W. (1990). The persistence of privacy: Autonomy and initiative in teachers' professional relations. *Teachers College Record* 91 (4): 509–536.

Lortie, D. (1975). *Schoolteacher*. Chicago: University of Chicago Press.

Louis, K.S. (1994). Beyond managed change. *School Effectiveness and School Improvement* 5 (1).

Louis, K.S. & Kruse, S.D. (1995). *Professionalism and community: Perspectives from urban schools*. Thousand Oaks: Corwin Press.

Louis, K.S. & Miles, M.B. (1991). *Improving the urban high school: What works and why*. New York: Teachers College Press.

Senge, P. (1990). *The fifth discipline*. New York: Doubleday.

Turnbull, D. (1994). Local knowledge and comparative scientific traditions. *Knowledge and Policy* 8: 29–54.

3

Informal Learning Communities and Their Effects

Miriam Ben-Peretz and Shifra Schonmann

The social organization of teachers in schools is framed by their common work-places and the norms of professional behavior that characterize these work-places. Largely because of such frame factors, the kinds of interactions in which teachers typically engage are limited. As Sarason (1996) points out:

> More often than not the teachers within a school do not feel them-selves to be part of a working or planning group. They may identify with each other in terms of role and place of work, and they may have a feeling of loyalty to each other and the school, but it is rare that they feel part of a working group that discusses, plans, and helps make educational decisions. (p. 141)

Feelings and behaviors of this sort are important to understand if schools are to become the learning organizations and professional communities that many people now believe is necessary for authentic and sustained school improvement (Louis, 1994; Talbert & McLaughlin, 1994).

In light of such norms and behaviors, how do strong professional communi-ties develop in schools? How can school cultures, often dominated by norms of privacy, be transformed into settings which value the sort of sharing and com-munal reflection so necessary to collective professional learning? In this chapter we explore the largely overlooked context of the teachers' lounge as a potential site for the enactment of professional community and the stimulation of collec-tive learning. In doing so, issues of culture, community and learning intersect in complex ways that turn out to be functional for the development of many teachers (but not all).

Framework

Our orientation to the study of teachers' lounges was framed by an interest in the concepts of professional community, organizational learning, and teacher collegiality and culture. These concepts are highly interdependent, and their manifestation in real schools is difficult to tease apart. In this section, we sketch in our understanding of these concepts, the nature of their interactions, and their anticipated meaning in the context of teachers' lounges.

Professional community

Talbert and McLaughlin (1994) propose that "professionalism among teachers is a product of social interaction and negotiation of norms within collegial work groups or networks" (p. 131). Beyond social interactions and negotiations of norms we view creation of communal knowledge and collective catharsis as central components in the development of professional communities. The teachers' lounge might be viewed as an ideal setting for the kind of interactions and negotiations to take place which lead to enhanced professionalism among teachers. The data of Talbert and McLaughlin reveal that:

> teachers who participate in strong professional communities within their subject area departments or other teacher networks have higher levels of professionalism, as measured in this study, than do teachers in less collegial settings. (pp. 142–143)

The prospects for enhancing professionalism are perceived by Talbert and McLaughlin to be determined locally as colleagues come to share standards for educational practice, including strong commitments to students and to their profession. In this view, local communities of teachers are the vehicles for enhanced professionalism in teaching.

According to Lieberman (1995):

> teachers must have opportunities to discuss, think about, try out, and hone out new practices. (p. 593)

Lieberman argues for making professional learning an integral part of the culture of schools. She suggests several ways of enabling teachers to do this in a school-based environment; for instance, by creating new structures such as problem-solving groups, or decision-making teams, and by creating a culture of inquiry as an on-going part of teaching. The movement, as she argued, is toward long-term, continuous learning in the context of school and classroom and with the support of colleagues.

It seems highly probable that teachers' lounges, as sites for interaction, provide the necessary conditions for the development of strong teacher networks, and the generation of communal knowledge about teaching. This knowledge might lead to improved teaching and learning.

Rosenholtz's (1989) work emphasizes the importance of teachers' collaboration in ensuring the academic success of their students. She states that:

> teachers' learning opportunities offer them a sense of ongoing challenge and continuous growth that makes greater mastery and control of their environment possible. (p. 7)

Moreover, she argues that teachers need less ego-endangering workplace circumstances in order to request and to offer professional advice.

> Norms of collegiality enable if not compel teachers to request and offer advice and assistance in helping their colleagues improve. We also find that, the greater teachers' opportunities for learning, the more their students tend to learn. (p. 7)

We perceive the teachers' lounge as potentially embodying conditions which promote teacher collaboration and mutual learning.

Events in teachers' lounges are communicative by their nature. They are social phenomena, structured in the context of time, place, and the involved participants, the teachers. An important term for understanding the social situation of lounges is Weber's (1968) term "social action." The importance of the term "social action" lies in the understanding that human actions are deliberate. Actions are not automatic or accidental, but express the ability of persons to grasp their environment and act in a reasonable and planned way, based on their interpretations. The opportunities for such deliberate interactions with peers provided by teachers' lounges are extremely important for counteracting the sense of being unsupported and alone. The teachers' lounge is the only unit in the school in which it is possible to examine events from the point of view of teachers being jointly together. Being in the teachers' lounge, as part of a collective, transforms teachers there into an entity which can be characterized in terms of the "lounge community." This community is conceived by its members, and by the students and their parents, as one group. Such "being together" is in contrast to the separate contexts of the cellular structure of classes from which teachers are coming and to which they are going to return with the ring of the bell.

We view the social activity in lounges as a cultural action in which teachers establish intellectual, emotional, and occupational roles. It is a social situation in which pure privacy does not exist, the teacher is forced to "take a role" and to play a part in a social situation. The blurring of boundaries between public and private spheres is based on an approach which emphasizes the duality in the nature of any social reality. Social facts are observable occurrences in such a social reality. They help to describe the social situation of teachers' lounges, as well as of classrooms. Shapiro and Ben-Eliezer (1991) explain a classroom situation in terms of social facts; i.e., students are sitting, the teacher is standing, the teacher asks questions, students answer, students who wish to speak raise their hands. The classroom is described as a well-defined place where activities planned beforehand are carried out. In the frame of these activities some of the participants are called "students." Their freedom of action is limited by a set of clearly defined rules which permit or prohibit certain behaviors.

Social facts, as Shapiro and Ben-Eliezer argue, shape the identity of classes which might be described as places in which teacher and students are partners in the same social entity. Classes differ from each other in the specific way their inhabitants are acting. According to Shapiro and Ben-Eliezer, some classes are characterized by bad manners in the acting out of social facts, like speaking without permission or making a lot of noise, and there are other, highly disciplined classes. Still, the notion of social facts applies to all classes. This concept of "social fact" might be applied as well to teachers' lounges: teachers come in, read the information presented on the notice boards, check their mailboxes, prepare coffee for themselves. Students are not allowed in. The ringing of the bell is the sign to stop or to begin any activity.

Like classrooms, teachers' lounges might differ in the manner in which their inhabitants express these social facts, potentially creating very different lounge climates. The social situation of lounges is a blend of observable events, social actions and social facts, in which one can participate. Hidden from the public eye, hard to define events are also part of the life in lounges, and are often the cause of individuals detaching themselves from the group, feeling uncomfortable and uneasy.

The idea of covert events leads us to think of the inner grammar of the teachers' lounges, which encompasses both overt and covert elements in the lounge. The inner grammar of the teachers' lounge includes both routines and unexpected events. It is important to note that the events in teachers' lounges, even in similar social situations, still have their own unique features which create the special climate of each lounge, reflecting the ethos of specific schools. It is necessary to determine what is the common ground and what is unique in the inner grammar of different lounges in order to derive an understanding of the significance of the social actions and the social facts within a particular lounge. Common and routine events in the lounge, such as providing information about a student, or an upcoming lecture, become part of the common basis for interpretation, and their meaning is easily shared among all participants in the social situation. Sometimes, however, the events in the teachers' lounge are uncommon, unique, and the individual interpretation of the situation is not part of the general, accepted basis for interpretation. The communicative actions in the teachers' lounge, which develop into routine patterns of behavior, unchanged over time, will be perceived here as one part of the inner grammar of teachers' lounges and will serve to define collegial norms.

Organizational learning

The orientation to our inquiry about collective or organizational learning in this study was framed by our interests in discourse communities, teachers as learners, and the concept of collective remembering.

Communities of discourse

Teachers' lounges hold the potential for fostering teacher development to the extent that they function as "communities of discourse". Such communities create a special kind of knowledge which originates in the professional conversations that may be held in the lounge, and which serves the everyday work of teachers. In the lounge the personal professional knowledge of teachers may be transformed into communal knowledge through the process of interaction and sharing between members of the community of discourse. These might be communities of subject matter domains (such as math teachers), or communities of teachers of children of the same age.

The potential discourse practiced in the teachers' lounge may be interpreted as the pursuit of further knowledge. It consists of both professional dialogue, and socially intimate dialogue. Throughout this discourse a new type of knowledge may be created, a form of knowledge that may not be created in other contexts. We can define this type of knowledge as "communal knowledge," a form of teachers' professional knowledge obtained through participating in events taking place within the teachers' lounge. Communal knowledge is based on the shared experience of teachers, and on the set of conventions established, in this case, in the teachers' lounge.

The term "shared experiences" has a double meaning. First, teachers share their classroom experiences in the lounge, and second, the interactions in the lounge are themselves a set of shared experiences which might have far-reaching implications for the practice of these teachers. Teachers' personal experiences, reflected on individually, constitute the personal wisdom of teachers. The interpersonal sharing and reflecting on professional experiences leads to the construction of a shared, communal wisdom of practice.

Interaction in the lounge is potentially different from the interaction between teachers in some other environments (e.g. private homes) since the lounge is a context of immediate practice. Walls and tables in lounges usually bear the signs of professional activity, announcements of meetings, or stacks of tests. Limitations of time between the rings of the bell, as well as the concentration of many persons in the limited physical space of the lounge, rarely allow for leisurely reflection. The intensity of such an environment seems likely to keep attention focused on pressing professional matters, and the formation of professional knowledge. This environment is suited to stimulating thinking and talking about matters in a highly concentrated form. Teachers in their lounges are in transition time, between periods of teaching in classrooms. Any knowledge shared with others, or generated in the community of discourse, might be perceived as serving practice immediately: it is most likely to be oriented toward the teaching of specific content areas, but may rely heavily, as well, on the affective dimensions of teaching, such as caring for individual students or feelings of joy, sadness, or disenchantment, which might accompany classroom teaching. The notion of creating communal knowledge embodies a view of teachers as engaged in ongoing learning.

Teachers as learners

Life in teachers' lounges also can be understood, after Gordon (1988), as text to be interpreted by members of a community. In order to understand the communicative process in the lounge one has to be aware that the role a teacher chooses to play there might differ from the role adopted when entering the classroom. In the lounge teachers interact with their peers and colleagues. The roles they choose to play may help to create an unwritten text code which is practiced in the lounge. Such a many-layered behavioral text contains messages that can be deciphered by the participants. Thus, teachers might arrive at a mutual understanding with their colleagues concerning the culture of the lounge and the school environment.

In their lounges, teachers might be conceived of as learners in a double sense. As inhabitants of the lounge, they learn to comprehend its text and to be more knowledgeable about its functions and conventions. Thus, teachers might gain understanding of norms of acting and reacting in this environment. On the other hand, teachers are likely to learn to decipher the professional aspects of life in school as participants in the generation of communal knowledge in the lounge.

We contend that the special circumstances of life in teachers' lounges may act, as well, as catalytic agents in the process of creating communal professional knowledge in the lounge. The lounge provides respite from the ongoing tensions of classroom events, and is also a haven for engaging in social interactions with one's peers. In the lounge, teachers act simultaneously as teachers and learners. Through telling their stories and sharing their experiences they stimulate and support each other in constructing and shaping professional knowledge. Telling stories is one way of creating and sharing knowledge, which becomes part of the communal knowledge of their school and is at the basis of its ethos.

Our study leads us to view the lounge as a dynamic system in which manifold voices interact. Thus, a balance is kept between creating communal school knowledge and maintaining the expressions of individual teachers. A major part of this system concerns processes of collective remembering.

Collective remembering

Learning from experience is a process of social construction, a process which may be understood not only from the two perspectives discussed above but also in terms of collective remembering. Middleton and Edwards (1990) emphasize the theme of knowledge sharing and using "as a form of collective memory in work settings" (p. 16). And, according to Orr (1990), the most notable characteristic of talk about work is its narrative structure:

> The stories in the community memory are produced in the real work
> of diagnosis and are further useful because their form prepares them
> for use in the next narrative creation of sense out of facts. (p. 186)

Beyond the construction and sharing of knowledge, the narrative form of community memory fulfills another important function, the shaping of the professional identity of members of the community. The process of collective remembering is crucial for the teaching profession, as teachers share their knowledge in the form of stories and narratives of events. Teachers' lounges might be an ideal setting for such collective remembering to take place. Ben- Peretz (1995) claims that:

> As one's individual professional identity is shaped through collective remembering, so does this process serve to establish the ethos of schools. Telling "our" stories and discussing their implications for action creates a common way of doing things in "our" school. The process of collective remembering becomes a channel for the induction of novice teachers to the culture of schools. (p. 140)

The shared professional knowledge of teachers determines how "we evaluate students," or how "one deals with classroom management problems in our school." Socially constructed communal knowledge is not static but highly dynamic and reconstructed continuously because of the constant flow of information, and the sharing of experiences which characterizes lounges.

Though the process of generating communal knowledge is part of the culture of all teachers' lounges, its specific features might differ greatly among various lounges and various groups of teachers. Teaching is often claimed to be a profession without ranking and inner hierarchy. To paraphrase Gertrude Stein, "a teacher is a teacher is a teacher." Yet, one can distinguish different groups of teachers in reference to their subject matter, their seniority, or their role in school. Teachers might enjoy different status according to these differentiations. Thus, teachers with more seniority might be considered to have a higher status in school and in the lounge. Their status might determine their role in the generation of communal knowledge. The personal practical and professional knowledge of senior teachers, or department chairs, might be perceived by their colleagues in the lounge as more valuable than the personal knowledge of novice teachers. This kind of hierarchy is dependent on the environment of teachers' lounges, and might be meaningless in other contexts. In meetings of teachers at each other's homes the special status of more senior teachers, or department chairs, might be much less important. Moreover, whenever high-status teachers, such as those with more seniority, do not participate in the give and take of professional knowledge in their community of discourse in lounges, they risk losing their status position.

There are likely to be different categories of interpersonal relations in the teachers' lounges, which might have an impact on the creation of communal knowledge and professional development. Events in the lounge, as witnessed in our interviews and observations, served to clarify this point, as related to interactions among peers, and among novice and senior teachers.

Collegiality and teachers' cultures
Teacher collegiality is considered to be a critical element of school cultures which foster collective learning. Such collegiality provides teachers with oppor-

tunities for developing instructional range, depth, and flexibility (Little, 1987). Interdependent collegiality sustains collaborative practices which, in turn, stimulate teachers to raise questions about their current practices, thereby fostering pedagogical innovation (Grimmet & Crehan, 1992). And yet, this kind of collegiality is rare in schools. Hargreaves (1989) argues that the prevailing context of teachers' work is not compatible with interdependent collegiality, and one might find a culture of "contrived collegiality," characterized by:

> a set of formal, specific bureaucratic procedures to increase the attention given to joint teacher planning and consultation. It can be seen in initiatives such as peer coaching, mentor teaching, joint planning in specially provided rooms, formally scheduled meetings and clear job descriptions, and training programmes for these in consultative roles. (p. 19)

Teachers' lounges are not likely to be much influenced by such bureaucratic procedures. The lounge is a place where teachers may enjoy private time in their own territory, a territory which represents the antithesis of contrived collegiality. Events in the lounge are usually unplanned, non-formal, and no bureaucratic regulations shape the ebb and flow of life in the lounge. By its nature the lounge is an environment in which the borders between the private and public spheres of teachers' lives are blurred. In this environment communities of discourse tend to develop and new, shared knowledge is generated.

Social cohesiveness and norms of collegiality enable teachers to create an atmosphere conducive to productive work. Teachers learn those parts of school life which are going to lead either to disagreement or to cooperative and enjoyable work. The teachers' lounge can be viewed as a shelter, a place of refuge from the complex, conflict-laden interaction between staff and students. The metaphor of "shelter" was suggested by the teachers themselves and was brought up constantly throughout our interviews with teachers. This figurative notion was expressed by Yona, an experienced high-school teacher:

> The teacher needs to liberate himself from the constant mask he is forced to wear; he has to "release steam" and behave naturally with friends. When the bell rings, the teacher, metaphorically speaking, is absorbed into the teachers' lounge, which becomes a shelter.

A basic assumption in this study was that teachers' lounges constitute unique territories in the educational environments of schools. By definition, the lounge belongs to the teachers and the activities that occur within its walls reflect the interconnectedness of teachers. This assumption is quite unlike Lortie's (1975) argument that "relationships with other adults do not stand at the heart of teachers' psychological world" (p. 187). Although Lortie was aware of the complex relationships among teachers, he did not examine the teachers' lounge, which we believe is a central site for deciphering these relationships. While trying to avoid oversimplification, Lortie, in contrast, argued that:

> The cellular form of school organization, and the attendant time and space ecology, puts interactions between teachers at the margin of their daily work. Individualism characterizes their socialization . . . It seems that teachers can work effectively without the active assistance of colleagues, since teacher–teacher interaction does not play a critical part in the work life of our respondents. (p. 192)

Lortie claimed that, being shaped by deeper commitments to students, the relationships teachers have with other adults are secondary, and derivative in nature. Goodlad and Klein (1970) made a similar claim:

> It would appear that teachers are very much alone in their work. It is not just a matter of being alone, all alone with children in a classroom cell, although this is a significant part of their aloneness. Rather, it is the feeling and in large measure the actuality—of not being supported by someone who knows about their work, who is sympathetic to it, who wants to help and, indeed, does help. This is, in part, an unhappy consequence of the inviolate status of the classroom and the assumed autonomy of the teacher in it. This aloneness becomes poignant in the face of problems which, clearly, cannot be solved by the individual teacher alone. (pp. 93–94)

We questioned the validity of these earlier claims, as applied to today's schools, and inquired about them in the context of the teachers' lounge.

Methods

The study was conducted in Israeli elementary, junior, and high schools. Teachers in Israel at all levels of schooling usually spend the breaks between classes in a common lounge, as there are no private spaces for them, nor is it customary to stay in their classrooms. The school day in Israel, at all levels, generally starts around 7 or 8 a.m. and finishes around 1 or 2 p. m. During this time there are three short breaks of 10 minutes each, and one longer break of 20–30 minutes. There is usually no lunch served at school.

Data for the study were collected as part of a larger study into the effects of teacher lounge environments on professional development of teachers. These data were of several sorts. Video recordings were made of five lounges, from different schools (two elementary schools, two junior schools, and one high school). Observation data were collected at twelve lounges when teachers were engaged in their daily affairs. Additionally, 19 teachers were interviewed, and 53 teacher wrote monologues about their lounges: "The lounge and I." These written monologues contribute to an awareness about the teachers' own feelings towards the lounges. Listening to the voices of teachers, and observing their daily life in the lounge, provided rich evidence about the meaning of lounges for teachers. We focused on relationships among teachers, and we observed interactions between teachers and principals, in order to gain insights into possible roles the lounge plays in the overall culture of the school.

Results and discussion

Professional community
Evidence from our study helps at least to illustrate how teachers sustain a sense of professional community through the teachers' lounge. These sustaining practices consist of efforts to maintain social cohesiveness in the face of it being threatened, providing social support for colleagues in difficult times, and allowing the lounge to serve as a forum for emotional catharsis when individual teachers need such a cathartic outlet.

Responding to threats to cohesion
In cases which threaten to weaken the social cohesiveness or sense of community of a lounge, one of the teachers might decide to take action. Nira's story is an example of such a situation. A conflict between two teachers caused increasing quarrels in the presence of the other teachers in the lounge. Only the bell released them from embarrassment and humiliation. Nira thought it was disgraceful and, regardless of who was responsible, she organized the teachers in the lounge to decide that the two teachers would not be able to enter the lounge until they apologized in front of their colleagues. One of the teachers did apologize willingly, because he felt uneasy about the whole situation, but the other refused. Gaps of misunderstanding increased, and this teacher found it unavoidable to leave the school at the end of the year.

Providing a cathartic outlet
The concept of catharsis, a purging of one's fears, anger, and frustration, can be useful for understanding one of the ways that the teachers' lounges contributed to professional community. According to Courtney (1987), catharsis is "the purgation of unwanted emotion through dramatic action" (p. 15). The cathartic nature of the teachers' lounge helped teachers cope with such feelings in the complex situations in which they were involved. Catharsis is a therapeutic need for teachers, as well as others, providing a constructive channel of release for their emotions. This helps relieve the social friction in teachers' lives. Since we are speaking about teachers as a community in the teachers' lounge, we define the catharsis there as a collective one, by its nature.

One can discern in catharsis two aspects of the very same phenomenon. Firstly, as a process of diminution, distribution, or division. Secondly, as a process of addition, of multiplicity or completion. The first aspect could be characterized as revealing a certain truth by deducting or removing the very parts that conceal the truth. The second could be characterized as a healing element which brings harmony within the unity of the contrasts of a situation. One can discern the duality in catharsis in different themes evolving in teachers' lounges.

The cathartic needs teachers expressed in the context of the teachers' lounges followed several themes. Teachers are accustomed to "grumble" in the teachers' lounge, to loosen up and to speak freely about everything that bothers them, to vent their frustrations (Keinan, 1994). When colleagues encourage and support the teacher's outburst, the cathartic process takes on the characteristic of a social support function of healing, leading the teacher to sense an abatement of negative feelings. If the teacher's outburst is rejected by his or her colleagues, the cathartic function is not fulfilled. Teachers speak of their successes, their failures, and their professional fears. Their talk might extend beyond the current affairs of school life and include issues such as the low income of teachers, problems with the teachers' organization, or the failure of the municipality to supply basic school needs. However, the cathartic themes in the teachers' lounge usually concerned the daily affairs of the school. The most common themes were: the insolence of a student, homework not being done, an undisciplined class, absenteeism of students, a dirty and messy classroom, a general failure in exams.

When a teacher enters the teachers' lounge he or she might "loosen up" and start a collective cathartic process among the other teachers. This process usually develops in the following way: the teacher enters the teachers' lounge, unloads a problem, and talks about what is bothering him/her. The theme is usually familiar to other teachers, who consequently react. Generally, owing to the fact that the problem must be dealt with immediately, the theme is taken up by a whole group of teachers. The theme is widely discussed and the process shared by members of the community. Thus the cathartic process becomes collective.

Two cases from our study, the case of Yardena and Shosh, illustrate the collective process of catharsis. Yardena entered the teachers' lounge in a state of near hysteria. She shut the door behind her, leaned on it so no one could come in, then burst into tears: "If Talmon [the vice principal, who is in charge of organizing the time-table] continues to make announcements to my class without my consent, I will simply quit teaching." In respond to Yardena's complaints, Bath-Ami responded and told the group that a certain exam assigned for that day had been cancelled without her even being consulted. When she entered the class with the prepared exams, she was informed by her students that "Talmon had cancelled the exams." Sarah then remarked cynically that teachers who let themselves be treated like rubbish were stupid. There was an immediate reaction by all the other teachers present in the teachers' lounge, many of whom had stories about the same theme. Collective catharsis reached its peak with raised voices, as teachers suggested ideas of resistance to the way the school was being managed. The collective involvement in the case served to relieve the anger and the teachers returned with spent and calmed feelings to their classes. Yardena's case can be viewed as an example of the healing aspect of catharsis due to the process of joint discussion of multiple instances of the same phenomenon.

Shosh's case also is illustrative of the collective cathartic effect of the lounge. "I've had enough! From now on I won't try to look for alternative teaching methods. I will simply lecture all the time! The method of chalk and talk is all they [the students] deserve, since they take advanatage of the liberty I've given them. From now on they won't have any choice." Shosh's emotional outburst was responded to by Dana, a young teacher who had been struggling to find a solution to the same problem for a long time and had been blaming herself for not succeeding. Since Shosh was regarded as an experienced, skilled, and creative teacher, her outburst helped Dana, who was then legitimately able to tell of the difficulties she had experienced that morning. Dana had divided the class into working groups; one group went to work in the school grounds and didn't return to the class before the bell ending the lesson had rung. Ruth said that the whole idea of dividing the class into working groups was worthless, thereby arousing Ettie's rage. Ettie was convinced that only bad teachers fail to deal with that method. A heated discussion on teaching methods enabled Dana to understand the complexity of the matter and overcome her feelings of frustration. This case exemplifies how catharsis can result from a better understanding of the complex nature of the problem, giving rise to emotions, and understanding developed in a social setting.

As soon as the teachers' feelings are accepted as legitimate, they become members of the collective team, "one of the family." Teachers say that the teachers' lounge is essential, one cannot do without it. The social function of catharsis in the teachers' lounge is perceived as necessary to the practice of teaching. It is possible to argue that, through collective catharsis, teachers cope better with the emotionally charged aspects of their work and/or find partners to share in the triumph and despair connected to that work.

Organizational learning
Evidence from our study concerning the collective learning that occurred in teachers' lounges is described in this section in terms of the interactions that occurred among peers, and between novices and senior teachers.

Interactions among peers
During breaks in the teachers' lounge, Debby sits close to the table of the English teachers, "where they regularly exchange comments over what is happening in their classes, what has succeeded or failed. Instructional material is exchanged along with the coffee." Debby confessed that sitting there gives her a sense of belonging, a place where her feelings of joy or despair can be expressed freely. She states: "My best friends sit here, and they understand me best." During the breaks, in Debby's teachers' lounge, teachers usually sit in the company of colleagues that teach the same subject, creating a regular pattern. She claims that this is only natural, since teaching has always been divided into subjects. She regards this division as most appropriate: "It's true that the teachers' lounge enables encounters with all kinds of teachers; however, it's better to sit near a 'regular' table and feel 'at home'." This was the picture drawn by Debby in her interview. Debbie has ten years of seniority, six of which she practiced at the same high school. Close and daily interactions among teachers who teach the same subject area provide ample opportunities for professional and social development.

In the teachers' lounge of another high school, political debates are a daily occurrence, perhaps due to the location of the school near the border, or because of the high political awareness of its population. The manner of sitting is not regular and fixed. The teachers form *ad hoc* groups as a result of the subjects being discussed. "The teachers' lounge is crowded, disgusting, and totally unaesthetic," claims Miriam, "yet a magic wand draws me in to participate in the discussions taking place there. Relationships that form within the teachers' lounge eventually grow into friendships beyond the daily work encounters." In this manner, lounges may be perceived as fostering the social development of teachers, the ability to relate to other teachers.

These examples suggest that teachers' lounges develop their own mode of discourse. The specific style of interactions is derived without any interference from a higher authority, but as a direct outcome of conventions and forms of discourse that create a subculture which is molded by the teachers and provide the basis for forming a collaborative culture serving teacher development and higher levels of professionalism. One teacher describes this process:

> The teachers' lounge is an important and vital place, mainly because it allows teachers to argue, express their opinions, to raise problems, and to search for solutions with one's colleagues. This enables teachers to change their views or their teaching strategies, following the deliberations in the lounge, when one finds out that the previous strategies are not appropriate for one's students.

This quote describes vividly how teachers develop professional ideas and actions. The lounge is perceived as a site for professional encounters. The discourse which develops there helps teachers solve problems in their practice. The immediacy of these professional interactions in the concrete context of teaching gives them their transformative power.

Interaction between novices and senior teachers

New teachers who have only recently become a part of the school staff have to adapt to the existing subcultures. Only after doing so will they feel at home in the lounge. The new teachers have to cope with the stress of new demands and have to prove themselves professionally. One of their first tasks usually is to communicate with the senior teachers in the school, win their trust, and possibly their friendship. Bearing in mind that each staff maintains its own norms, connected with subject-matter subcultures within the teachers' lounge, the new teacher soon realizes that it is necessary to get acquainted with the process we have defined as the creation of communal knowledge. The new teacher attempting to minimize the inevitable gap between their former learning and the learning going on in the lounge might be very hesitant. Overcoming these gaps is part of teachers' induction into practice through participation in the community of discourse in the teachers' lounge, among colleagues. This is a process by which the new teacher's future competence might be shaped.

Young teachers tend to feel like outsiders, as Ronnie, a first-year teacher, commented:

> I'm uncomfortable and feel that I'm being judged all the time. Perhaps it's a matter of seniority and in the course of time I will get used to it.

Lisa refers to the bad feeling that accompanied her first months of teaching:

> Senior teachers are, as a rule, snobbish. When at last they do inquire whether you are getting along, they don't even bother to listen to the answer.

Jane, another teacher who looks particularly young, aggressively declares that:

> Senior teachers tend to pressurize me. They don't create a pleasant atmosphere. They simply don't assist or anticipate. They always test you and your looks. Every day someone remarks that I look like a teenager myself. In fact, I wouldn't mind if a few of the teachers retired early.

Whenever novice teachers are confronted with teachers with seniority, a problem of status tends to arise. Adam, in his first year of teaching, states:

> The mentality which governs the teachers' lounge disgusts me. I feel very frustrated, surrounded by this atmosphere in which I'm slowly sinking and which does not reflect well on any of the senior teachers.

Adam defined a strict line between male teachers and female teachers, yet he also refers to the seniors as a monolithic group defeat. He sees ugliness around him and feels repelled, which makes it extremely difficult for him to become part of their community of discourse.

> I would like to make them laugh, to inquire what was so tragic, yet I am afraid they will consider me a "nuisance."

At first Adam tried to establish contact with a number of the senior teachers by informing them about his work in class. He had expected encouragement but the general response was: "Never mind, it will be easier in the days to come," or "I've been through it myself." Adam had started his working life motivated by concepts of innovation and change, and an autonomous perception of teaching. However, when confronted with the indifference of some senior teachers, who were burned out, he sensed they had constructed a solid wall around themselves which was growing high and was conceived by him as dangerous to his own professional development. Adam faces a dilemma: whether to extend his efforts to establish contact with the senior staff, or to break off from them. Most new teachers long for some special care and personal relations, the way a new student in class struggles to determine his/her status within the group.

Integration within the teachers' lounge has a great effect on the young teacher's decision about whether he/she will remain a teacher. An example of this can be seen in the case of a literature teacher who admitted she quit teaching at one school because of the teachers' lounge, which she regarded as a "hornet's nest." She found her colleagues there troublesome, contemptible, and all too happy to gossip. In contrast to these descriptions, there is evidence of support and assistance. Some teachers' lounges have a round table in the center. In those lounges, even if there are defined groups, the atmosphere is one of acceptance and cooperation. In these circumstances there are manifold opportunities for the personal and social development of teachers.

Yet, the general picture clearly indicates that there is a distinction between novice and senior teachers. Naomi, who teaches at a high school, confessed during her interview that:

> Most of the senior teachers are very old. There exists a sense of reverence towards them, partly due to the fact that they are a strongly consolidated group whose code of language is unique to them, thus only a few of us young teachers succeed in interacting with them.

A possible explanation for the experiences of these novice teachers is based on the notion of the origination of communal knowledge in the lounge. This knowledge has usually grown over many years, it might be expressed in "code words," and it is extremely difficult for newcomers to interpret or to join in its ongoing development. Moreover, such knowledge cannot be acquired passively; it has to be constructed through active engagement in the discourse, and novice teachers might be shut out or might act only on the fringe of this process. This interpretation is reinforced from the opposite point of view. Noga, a senior teacher, admits that:

> They are too young and I feel like "the old billy-goat." With my experience, what can a younger teacher possibly offer me. Thus cooperation between us can only be made possible on the basis of ourselves acting as mentors.

Thus the novice teacher is effectively made to stay outside the community of discourse among equals. In the teachers' lounge, we are apt to find complex encounters, imbued with dilemmas, between two biological and professional generations, each of which criticizes the other, tests the other. Whether it is deliberate or not, many senior teachers create a wall to protect their privacy, and breaking through this barrier is enormously difficult for novice teachers. Yet, the interaction between senior and novice teachers seems to be an integral component for mutual learning to go on.

In the everyday lives of teachers there is always the danger of loneliness and despair. Only strong professional communities have the power to counteract these feelings.

Up to this point we have focused mainly on lounges as sites for creating communal knowledge. We turn now to the social aspects of the lounge and to the notion of collegiality.

Collegiality and teachers' culture

The term collegiality has remained conceptually amorphous and ideologically sanguine, as Little (1990) claims:

> Advocates have imbued it with a sense of virtue—the expectation that any interaction that breaks the isolation of teachers will contribute in some fashion to the knowledge, skill, judgment or commitment that individuals bring to their work and will enhance the collective capacity of groups or institutions. (p. 509)

Norms are perceived as shaping our social actions and the social facts of a situation, as well as determining how we interpret these actions. The norms of teachers' lounges are created during the daily interaction between teachers. Norms which concern ways of speaking, of dressing, and of conducting routine activities might be perceived as "collegial norms." People who have a group relationship tend to reach agreement in areas relevant to the interactions in this group. Norms can be implicit, and hard to identify. The more explicit the norms, the easier it is for new members to enter a group and adapt to its ways. The view of norms as unwritten rules and as expectations to behave in a certain manner provides an explanatory framework for some of the dilemmas, dramas, and tensions which arise in lounges, examples of which are presented in the following sections.

For example, according to the normative perception acceptable in teachers' lounges, students are not allowed to enter this area. When students do enter the lounge because a teacher invites them, important norms come into conflict and a dilemma is created. On one hand, the "collegial norms" of the lounge forbid student entry. On the other hand, assisting students is an important school norm governing the role of teachers. Dilemmas in the lounge can be caused by contrary expectations for behavior. For instance, in the lounge teachers might all talk at the same time, while in their classes a strict order of speaking might be adhered to. Some teachers might find this situation highly stressful, reacting verbally and trying to stop the racket.

Life in teachers' lounges reflects the culture of collegial norms, and it is important for inhabitants of this culture to understand the nature of these norms. Two events exemplify the role of norms of collegiality at play in the social situation of teachers' lounges.

Giving up a free day

In one teachers' lounge in a senior high school, a group of six English teachers were very close to each other. They used to eat breakfast together each Tuesday. That was their way of showing cohesiveness as a professional group. It became a ritual which involved bringing coffee and cake, telling jokes, and creating their own unique atmosphere. One morning the principal told one of the teachers that her schedule had been changed so in the future Tuesday would be her day off. The teacher told the principal that it was impossible for her to change days because at that time the English teachers meet and discuss their work in an informal meeting and she could not allow herself not to participate in it. She was willing to give up her day off for the sake of not withdrawing from the group of which she was a member.

Breaking the norm
Collegial norms might be considered by some members of the group to be most unpleasant. Eating in the teachers' lounge is one of the more obvious norms that exist. Teachers' routines—that is their choice of companions, the way they prepare their coffee—all reflect these norms. Tammi, for example, had many negative experiences to impart about the teachers' lounge. She told us that she had tried to bring flowerpots to decorate the room but no one had paid any attention. She became upset with the way everyone left their dirty glasses of coffee, breadcrumbs, and ashtrays on the table. Above all she was upset with the fact that she could not change these norms of behavior. She stopped entering the teachers' lounge, except to take her mail or read the announcements. However, since she had broken the norm of being part of the group, she felt that she could not stay in the same school. At the end of the year she moved to another school.

Misunderstandings, conflicts, and unpleasantness in the lounge can be explained on the basis of cultural and individual differences concerning the norms of using time and space (Hall, 1959). Thus, for example, a new teacher might enter the lounge and sit on a free chair without knowing that this space is usually occupied by one of the senior teachers. The new teacher might feel uneasy, without understanding the "hostility" around him or her.

Another common norm concerns the use of time in the lounge. The ringing of the bell after the break is a sign to return to classrooms, but teachers might tend to delay their exit from the lounge. Teachers who leave the lounge with the first sound of the bell break a collegial norm, and may become the object of jokes, or even hostile remarks.

Collegial norms of accepted behavior in the lounge are usually shared by faculty and students. Thus, students know whether the lounge is off-bounds or not, and when and under what circumstances they are allowed to enter it. Any unusual behavior tends to create tension. The following event is an example of such a situation.

Mary, a 17-year-old student, entered the lounge one day, and kissed and hugged her teacher. The other teachers who were there were unable to accept this as a normative way of behaving and revealed their dissatisfaction. They did not know that Mary had been seriously ill. She had returned to school after being treated for cancer for a long period. Esther, the teacher, was extremely pleased to see Mary and showed her delight. She could not tell her friends the reasons for the affectionate behavior because Mary's family had asked for the illness to be kept a secret.

Deciphering the nature of collegial norms provides insights into the life in teachers' lounges, and might have implications for teacher education.

Norms of "a good colleague"

Beyond the norms exemplified above, norms which determine the kind of behavior which is acceptable in the lounge, there are norms concerning the behavior of a "good colleague," "someone who makes professional lives easier."

From our interviews with teachers, it became clear that *firgun* is the central element in the concept of a "good colleague." *Firgun* is a very specific term in Hebrew slang for expressing lack of envy for the success of others. In teaching, this means the kind of relationship which serves to advance a good professional reputation. The "good colleague" is one who is not jealous of the successes of colleagues but, on the contrary, helps to further their reputation. This matter is of great importance. According to our interviews, this is considered by teachers to be more important than helping to prepare materials for a lesson, or helping to solve a problem. *Firgun* is central and basic in the perception of a "good colleague." How can we understand this phenomenon? *Firgun* is a process in which one praises a colleague and gives him/her one's full support. It is a social behavior which influences professional work. Teachers' work is influenced by the way that teachers advance each others' reputation. This is one factor in creation of social cohesion in the community of teachers. Social cohesion— *gibush*—is of great value in the life of Israeli teachers. It can be understood in the framework of Katriel and Nesher's (1986) and Katriel's (1991) works which studied the *gibush* in Israeli schools.

They found that one of the major concerns of every teacher is to create a cohesive atmosphere in the class. The same perception of good and cohesive interpersonal relationships is transferred into the culture of lounges, and shapes the norm of a "good colleague." Though the concepts of *firgun* and *gibush* are terms which emerged in the Israeli culture, they might be meaningful in other cultures of teaching as well, as they relate to some of the basic concerns of the profession. The main point we wish to make is that teachers are part of a professional community, even when they are in the confines of their classrooms. Many care about the fortunes of other teachers and are concerned about their own professional image.

It is important to note that the "good colleague" is not necessarily a unidimensional concept. Different teachers might create their own notion of "good colleagues." Still, it seems that, in the Israeli culture, features of mutual support, leading to social cohesiveness, are important qualities of "good colleagues" in the teachers' lounge.

Conclusion

Our observational evidence collected in teachers' lounges has illustrated just how interdependent are teachers' senses of professional community and the collective learning that occurs in that community. This evidence has also demonstrated the important contribution made to both collective learning and a sense of professional community by a teaching culture, within the school, characterized by strong norms of collegiality.

Viewed from the perspective taken in this study, teachers typically are well integrated members of a social organization, not the lonely individuals lacking in contact with their peers which have been portrayed by many in the past. And from this perspective, the teachers' lounge is a natural site for the transformation of these social organizations into active professional learning communities. Interactions of various sorts within these communities produce communal knowledge development and allow for collective catharsis, two important elements in this transformation process. The notion of communal knowledge and its development provide a framework for understanding teachers as simultaneously teaching and learning in their lounge. Collective catharsis provides opportunities for solving dilemmas and for reducing tensions in the workplace. In this way, catharsis promotes social cohesiveness, the basis for sustaining a learning community of teachers.

References

Ben-Peretz, M. (1995). *Learning from experience: Memory and the teacher's account of teaching*. Albany, NY: SUNY Press.

Boyce, S.N. (1987). *Welcome to the theatre*. Chicago: Nelson Hall.

Courtney, R. (1987). *Dictionary of developmental drama*. Illinois: Charles C. Thomas.

Goodlad, J.I. & Klein, F. (1970). *Behind the classroom door*. Worthington, OH: James.

Gordon, D. (1988). Education as text: The varieties of educational hiddenness. *Curriculum Inquiry* 18 (4): 425–449.

Grimmet, P.P. & Crehan, E.P. (1992). The nature of collegiality in teacher development: The case of clinical supervision. In A. Hargreaves & M.G. Fullan (eds.), *Understanding teacher development* (pp. 56–85). New York: Teachers College Press.

Hall, E.T. (1959). *The silent language*. New York: Garden City.

Hargreaves, A. (1989). *Contrived collegiality and the culture of teaching*. Paper presented at the annual meeting of the Canadian Society for the Study of Education, Quebec City.

Jackson, P. (1968). *Life in classrooms*. New York: Holt, Rinehart & Winston.

Katriel, T. (1991). *Communal webs*. Albany, NY: SUNY Press.

Katriel, T. & Nesher, P. (1986). Gibush: The rhetoric cohesion in Israeli school culture. *Comparative Education Review* 30 (2): 216–232.

Keinan, A. (1994). *The staffroom: Observing the professional culture of teachers*. Brock Field: Ashgall Publishing.

Lieberman, A. (1995). Practices that support teacher development. *Phi Delta Kappan* 76 (8): 591–600.

Light, D. (1980). *Becoming psychiatrists: The professional transformation of self*. New York: W.W. Norton.

Little, J.W. (1982). Norms of collegiality and experimentation: Workplace conditions of school success. *American Education Research Journal* 19 (3): 325–340.

Little, J.W. (1987). Teachers as colleagues. In V.R. Kohler (ed.), *Educators' handbook: A research perspective* (pp. 491–548). New York: Longman.

Little, J.W. (1990). The persistence of privacy: Autonomy and initiative in teachers' professional relations. *Teachers College Record* 91 (4): 509–536.

Lortie, D. (1975). *Schoolteacher*. Chicago: University of Chicago Press.

Louis, K.S. (1994). Beyond "managed change": Rethinking how schools improve. *School Effectiveness and School Improvement 5* (1): 2–24.

Middleton, D. & Edwards, D. (eds.) (1990). *Collective remembering*. London: Sage Publications.

Orr, J.E. (1990). Sharing knowledge celebrating identity: Community memory in a service culture. In D. Middleton & D. Edwards (eds.), *Collective remembering* (pp. 169–189). London: Sage Publications.

Rosenholtz, S.J. (1989). *Teachers' workplace: The social organization of schools*. New York and London: Teachers College Press.

Sarason, S.B. (1996). *Revisiting the culture of the school and the problem of change*. New York and London: Teachers College Press.

Shapiro, Y. & Ben-Eliezer, U. (1991). *Elements of sociology*. Tel-Aviv: Am Oved Publishers. (in Hebrew)

Talbert, J.E. & McLaughlin, M.W. (1994). Teacher professionalism in local school contexts. *American Journal of Education* 102: 123–153.

Weber, M. (1968). *Economy and society*. New York: Bedminster Press.

4

Leadership and other Conditions which Foster Organizational Learning in Schools

Kenneth Leithwood, Doris Jantzi, and Rosanne Steinbach

Like members of most contemporary organizations, educators are swamped with arguments for why they must change. Futurists point to current trends in society-at-large that portend enormous consequences for the design of future schools (e.g., Naisbitt & Aburdene, 1985). Reformers offer images of what future schools might look like were they to respond seriously to these trends (e.g., Dixon, 1992; Perkins, 1992; Schlecty, 1990). In developing a vision of schools capable of responding effectively to the challenge to change, two closely linked problems inherent in the positions adopted by these futurists and reformers must be solved. One problem is the risky business of predicting the future social and economic consequences of present trends; the other is the improbability of accurately and precisely specifying the characteristics of schools adapting to such consequences.

Envisioning future schools as learning organizations does not require exceptional accuracy in predicting consequences for the future of current trends. That schools will continue to face a steady stream of novel problems and ambitious demands is the only prediction required. These demands and problems most certainly will generate considerable pressure to learn new and more effective ways of doing business. This is a pretty safe bet. Indeed, it is a bet that a great many non-school organizations are prepared to make as they attempt to reinvent themselves. Vivid testimony to this claim is to be found in the remarkable following enjoyed by some of the more recent, popular accounts of learning organization (Senge, 1990; Watkins & Marsick, 1993). And while the nature of this following raises the specter of a quickly passing fad, the long-standing and distinguished literature on organizational learning within the domain of organizational theory argues otherwise (Argyris & Schön, 1978; Hedberg, 1981; Levitt & March, 1988).

The study on which this chapter is based was part of a five-year longitudinal study of policy implementation in the Canadian province of British Columbia (Leithwood et al., 1990, 1991, 1993a, 1993b, 1994). Since 1989, the government in that province has been pursuing initiatives originating from a Royal Commission report aimed at the comprehensive restructuring of schools. In the British Columbia context, "restructuring" encompasses changes in such aspects of schooling as curriculum, assessment, attention to diverse student needs, classroom and school organization, teacher development, the role of school leaders, relations with parents, and links with the wider community and with post-secondary educational institutions (British Columbia Ministry of Education, 1989).

Among the objectives in all phases of this research has been an attempt to account for variation among schools in the nature of their responses to these central policy initiatives, and the degree to which their responses have been productive. In some phases of this research, "productive" has been defined as a function of the extent to which policy initiatives have been implemented (Leithwood et al., 1990, 1991); in other phases, "productive" has been judged, more fundamentally, in terms of student outcomes (achievement, participation, and identification) (Leithwood et al., 1994). Results of previous phases of the research increasingly have directed attention toward individual and collective learning processes ("organizational learning" or OL as it will be referred to subsequently) as explanations for variation in the productivity of school responses. School leadership has emerged as an important explanation for variation in OL. Accordingly, the phase of the study described in this chapter was designed to explore more fully the causes and consequences of OL in schools and to discover those leadership practices which contribute to such learning.

Subsequent sections of the chapter describe the framework used to guide our inquiry about OL in schools, briefly summarize our research methods, and provide a synopsis of our results. These results focus mainly on the conditions (including leadership) found in schools which foster organizational learning. Creating these conditions ought to be the immediate aim of those wishing to enhance the individual and collective capacities of school staffs.

Framework and research questions

Fiol and Lyles (1985) suggest that organizational learning means "the process of improving actions through better knowledge and understanding" (p. 203). A learning organization is "a group of people pursuing common purposes (individual purposes as well) with a collective commitment to regularly weighing the value of those purposes, modifying them when that makes sense, and continuously developing more effective and efficient ways of accomplishing those purposes" (Leithwood & Aitken, 1995, p. 63).

An extensive literature on OL in non-school organizations (reviewed in Cousins, 1996; Leithwood & Aitken, 1995) was used to develop the framework for the study serving as the basis for this chapter. This framework consists of five sets of ideas and their relationships which, together, encompass an explanation for how and why OL occurs and what are its consequences. Research questions were keyed to each of these sets of ideas.

Stimulus for learning

OL is assumed to be prompted by some felt need (e.g. to respond to the call for implementing a new policy), or perception of a problem, prompted from inside or outside the school, that leads to a collective search for a solution. This study asked: What sorts of internal dispositions (on the part of individuals) or external events trigger organizational learning? Are policy initiatives by the Province and by districts among these triggering events? And how do such "official" initiatives compare with other types of initiatives in their power to stimulate OL?

Organizational learning processes

These are activities engaged in by individuals and groups within the school (e.g. informal discussion of new ideas; personal reading) to make sense of their environment and to master the challenges posed by that environment, and those mechanisms (e.g. workshops, staff meetings) used by organizational members for such sense-making and problem-solving. About these activities, we asked: What individual and collective processes account for OL? How can collective and individual learning processes be distinguished?

Out-of-school conditions

Included within the meaning of out-of-school conditions are initiatives taken by those outside the school (e.g. Ministry personnel, district staff), or conditions which exist outside the school (e.g. economic health of the community) that influence conditions and initiatives inside the school. This study focused only on those out-of-school conditions created by the Ministry of Education, the local school community, and the school district. We asked: What sorts of conditions outside of schools have a bearing on OL in schools? In particular, what is it about school districts, local school communities, and the Ministry of Education that fosters or inhibits OL in schools? What would be the characteristics of such an "external environment" which unambiguously nourished the development of schools as learning organizations?

School conditions

These are initiatives taken by those in the school, or conditions prevailing in the school, which either foster or inhibit organizational learning. In this study, such initiatives and conditions were associated with the school's mission and vision, school culture, decision-making structures, strategies used for change, and the nature of school policies along with the availability and distribution of resources. The study asked: What do schools look like when they are behaving like learning organizations? Specifically, what is it about a school's vision, culture, structure, strategies, and policies and resources which gives rise to or detracts from OL?

School leadership

Defining our meaning of school leadership were the practices of those in formal administrative roles that help determine the direction of change in the school and influence implementation of that change. While leadership is encompassed in our meaning of *school conditions*, it was treated separately, because of its probable effects on other school conditions as well as on OL, and because of the special interest it held for us. Research on OL in non-school organizations suggests that administrative leadership is an especially powerful influence on OL, both directly and indirectly. What sorts of leadership practices on the part of school administrators, we asked, contribute significantly to OL and to the conditions which foster OL?

Outcomes

The individual and collective understandings, skills, commitments, and new practices resulting from organizational learning on the part of school staffs were encompassed by our meaning of "outcomes." To be worth continuing attention, OL must result in something consequential for schools. Does it? What are the consequences or outcomes? Specifically, what individual and collective understandings, skills, commitments, and overt practices result from OL in schools? These outcomes are assumed to mediate the effects of the school on student growth.

The primary focus of this chapter is on leadership and other school conditions which foster OL. Nevertheless, results of the study concerning other aspects of our framework are also described as a means of better understanding the context within which leadership and other school conditions influence OL.

Methods

The primary source of data for the study on which the chapter is based were semi-structured interviews conducted in six schools located in four districts in British Columbia, Canada. These schools were selected as promising sites of organizational learning from two sources of evidence. One of these sources was the data available about the school as a result of its participation in one or more earlier phases of this research (four schools). Another source was a school's reputation among two or more district staff as a school making substantial progress toward restructuring (two schools), something akin to Rosenholtz's (1989) "moving schools." The cases also were selected to represent a broad spectrum of potential school organizations: one primary, one elementary, a junior secondary, two secondary, and one senior secondary school.

Principals in each school were asked to nominate up to 12 teachers who would be willing to be interviewed; nominees were to be broadly representative of the staff, with differences in curricular areas taught, years of experience, and gender, reflecting the variety of experience and expertise within the school. A total of 72 teachers and 6 principals were interviewed for this study; teacher interviews took about 50 minutes and the principals about 90 minutes.

Interview data were collected using an instrument which consisted of 28 questions about all components in the conceptual framework. These interviews were tape recorded, transcribed, and coded according to the framework. Analysis of the data included identifying the associations made explicitly by teachers among the variables in the framework. Remaining sections of the chapter describe what was learned from the teacher interviews concerning each of the sets of questions associated with the study's framework. Results in response to questions of leadership are introduced last and described more fully than other results.

Events and dispositions which "trigger" OL

> In one sense we're having to change to meet our new clients. Some of these kids that are now coming up into grade 9 had been at a middle school where some of the facets of the Year 2000 have been in place. For example, student conferencing, portfolios, students who have been involved with evaluating themselves, students who have had more control over what they are studying. While those kids are now only in grade 9, I think there's been a gradual movement to where some of the staff are starting to think about it more seriously.

A total of nine "external" and "internal" stimuli for individual and organizational learning were mentioned by many teachers. Prominent among external stimuli were "official" sources including new Ministry programs, new programs being implemented in one's school, encouragement from administrators to implement new programs, and district policy initiatives. Demograghic changes in the student population was the remaining external source. Internal sources of OL included a desire to improve one's practices, to do what is best for students, to move in the same direction as colleagues, and a belief that new programs were compatible with one's own professional goals and preferred teaching styles.

Teachers explicitly associated these stimuli with OL in all six schools and about the same number of stimuli were associated with collective as with individual learning. However, that pattern varied among schools. As these results suggest, then, quite a few things have the potential to stimulate OL and schools appear to vary in their sensitivity to these stimuli. This may well be a function of their missions and visions; some were more open to Ministry initiatives than others, for example. Some school cultures also may foster an openness to ideas from other schools or from one's own colleagues. OL can be stimulated by relatively everyday events: ongoing attempts at incremental improvement and the like. It does not require a crisis, as some of the literature from non-school organizations (e.g., Watkins & Marsick, 1993) appears to suggest.

Out-of-school conditions which influence OL

> This district has been big on kind of giving teachers new tools that
> have come out over the last few years. They spent thousands of dol-
> lars, probably three or four years ago, with many in-service sessions
> that anybody was welcome to. I think this district has done a fairly
> good job of helping teachers to change in their teaching methods.

Three sets of factors are identified in the framework as likely to influence
learning, either directly or indirectly (through their influence on school conditions): the school district, the local school community, and the Ministry of
Education.

District conditions
Little distinction was evident in the data concerning district conditions between
conditions which effected individual as distinct from collective learning. The *missions and visions* of school districts were potentially fruitful sources of learning for
school staffs. But, to realize this potential, such visions had to be well-understood,
meaningful, and accessible: they needed to engender a sense of commitment on the
part of school staffs. When these conditions were met, and when district visions
acknowledged the need for continuous professional growth, teachers and administrators used the visions as starting points and frameworks for envisioning more
specific futures for their own schools; in effect, establishing the long-term goals for
their own professional learning. Widely shared district missions and visions, furthermore, sometimes provided filters for screening and evaluating the salience of
external demands for change. Also, they served as non-prescriptive clues about
which initiatives taken by schools would be valued and supported by district personnel. It should be noted, however, that teachers identified district mission and
vision as the weakest of the district conditions influencing OL.

"Collaborative and harmonious" captures much of what was considered to
be important about district *cultures* when they contributed to OL (they were of
moderate influence). Rather than a "we–they" attitude, perceived to promote
hostility and resistance toward district initiatives, learning appears to have been
fostered by a shared sense of district community. This sense of community was
more likely when there was interaction with other schools (e.g. feeder schools),
and when disagreements in the district were settled in ways perceived to be
"professional." District cultures fostered OL also when the need for continuous
change was accepted, and when new initiatives clearly built on previous work
rather than being discontinuous with such work.

District *structures,* one of the strongest influences according to teachers, fostered OL when they provided ample opportunity for school-based staff to participate in shaping both district and school-level decisions. Participation in
district decisions teaches those involved about the wider issues faced by the district and those influences not readily evident in schools that are, nevertheless,
germane to district decisions. Considerable delegation of decision-making to
schools (possibly through site-based management) enhanced opportunities for

improving the collective problem-solving capacities of staff. Such decision-making also permitted staff to create solutions which were sensitive to important aspects of the school's context. Evidence suggests that multiple forums for participation in district decision-making were helpful.

To foster learning, it was perceived to be useful for districts to use many different *strategies* for reaching out to schools—through newsletters, workshops, informal lines of communication, and the like: teachers viewed these as moderate influences on OL. Especially influential, according to teachers, were strategies with teaching as their explicit purpose: workshops and mentoring programs were identified as examples by interviewees, as were specific change initiatives designed to assist in achieving district goals and priorities. Strategies which buffered schools from excessive turbulence or pressure from the community also were identified as helpful for learning.

District *policies and resources* were the stongest set of district conditions influencing OL. Promoting both individual and collective learning was the provision of release time for planning and for professional development, especially when these resources could be used in flexible ways. Access to special expertise or "technical assistance" in the form of consultants and lead teachers, for example, also was claimed to foster learning, although teachers reported that such resources were, by now, quite scarce ("in the past" they had been quite useful). One means identified for creating a critical mass of expertise about a focus within the school from which others could learn was to ensure that more than one participant from a school attended in-service events. In districts which had professional development libraries or central resource centres, teachers cited them as significant aids to their professional learning.

Community conditions

Interviews with teachers about community conditions fostering OL suggested that parental agreement with the school's direction and practices, an atmosphere that welcomed parents into the school, and active participation of parents in the school were important. Several of the six case schools were experiencing a rapid influx of students from different countries and cultures. These demographic changes in the student population challenged staffs to alter their programs, to find resources for ESL instruction, and to learn about the consequences of students' backgrounds for classroom practices. Changes in the socio-economic status of the student population required staffs to learn more about resources available to students and families from social service agencies and how to assist students in gaining access to these resources.

Ministry conditions

Financial support along with curriculum and other resource documents were identified as aids to learning provided by the Ministry. The Ministry also fostered learning through its sponsorship of an action research project in which a number of interviewees had participated. Coherent policies, consistent approaches to their implementation, and sustained commitment to those policies were identified as important conditions for OL in schools that the Ministry had *not* consistently provided, according to those interviewed.

Of the three sets of out-of-school conditions directly or indirectly associated with OL, district conditions clearly predominated; Ministry conditions were identified many fewer times and community conditions rarely. Among district conditions, greatest influence seemed to be exercised through district policies and resources, especially professional development resources. There was considerable variation among schools in the district, community, and Ministry conditions to which they were sensitive.

School conditions fostering organizational learning

> Informally, we do a lot of sharing in this school. The two teachers who are really working with the computer lab have been very generous in sharing their time and expertise, and getting us in there and getting hands-on, answering as many questions as they can, and making it easy for us as possible to work with the children. The PE people have shared, anybody who's been away at a workshop and picked up something that they think the rest of us would enjoy certainly have shared the information. And I think just generally people who go away to things and come back share the materials around. We don't tend to have closed doors in this school, which is really nice . . . we're all in it together.

Interviews with teachers identified conditions within schools which they believed either fostered or inhibited some form of individual or collective learning on their part (Table 4.1). As in the case of *district* vision and mission, school vision and mission was associated with OL when it was clear, accessible and widely shared by staff. To have this association, school vision had to be perceived by teachers as meaningful; it also had to be pervasive in conversations and decision-making throughout the school.

Also paralleling conditions at the district level, school *cultures* fostered learning when they were collaborative and collegial. Norms of mutual support among teachers, respect for colleagues' ideas, and a willingness to take risks in attempting new practices were all aspects of culture that teachers associated with their own learning. Some teachers indicated that receiving honest, candid feedback from their colleagues was an important factor in their learning. Teachers' commitments to their own learning appeared to be reinforced by shared celebrations of successes by staff and a strong focus on the needs of all students. Collaborative and collegial cultures resulted in informal sharing of ideas and materials among teachers which fostered OL, especially when continuous professional growth was a widely shared norm among staff.

For the most part, school *structures* believed to support professional learning were those which allowed for greater participation in decision-making by teachers. Such structures included: brief weekly planning meetings; frequent and often informal problem-solving sessions; regularly scheduled professional development time in school; and common preparation periods for teachers who needed to work together. Other structures also associated with learning were the cross-department appointment of teachers and team teaching. When decisions were made by staff through consensus—something easier to do in smaller schools—more learning was believed to occur. The physical space of schools had some bearing on teachers' learning, when it either encouraged or discouraged closer physical proximity of staff.

Clarifying short-term goals for improvement, and establishing personal, professional growth goals, were cited by teachers as school *strategies* that aided in their learning. This learning was further assisted when school goals and priorities were kept current through periodic review and revision and when there were well-designed processes for implementing those specific program initiatives in order to accomplish such goals and priorities. Schools fostered OL when they were able to establish a restricted, manageable number of priorities for action and when there was follow-through on plans for such action.

Teachers reported that sufficient *resources* to support essential professional development in aid of their initiatives was a decided boost to their learning. Within their own schools, teachers used colleagues as professional development resources, along with professional libraries and any professional readings that were circulated among staff. Access to rich curriculum resources and to computer facilities aided teachers' learning, in their view, as did access to technical assistance (consultants, etc.) for implementing new practices. Teachers also noted that access to community facilities helped them to learn.

Culture was the only category of school condition referred to by teachers, at least once on average, as explicitly related to collective learning. Least likely to be associated with learning was school *vision*—with one notable exception: in this school there was consensus about the vision among staff, at least in part due to emphasis having recently been given to creating a common vision. That process, teachers reported, did influence their learning. For similar reasons, *strategy* was associated with learning in a school engaged in systematic and authentic goal setting in a way that fostered professional development, according to teachers.

Surprisingly, teachers made few explicit associations between school *resources* and either collective or individual learning. This could be interpreted as evidence of the current lack of resources to support teacher learning (there were some negative associations between resources and OL). But teachers in the school for whom professional learning appeared to be an ongoing priority also made no explicit associations between resources and learning. On the other hand, another school recently had received extra funding for its involvement in the school accreditation process (a school evaluation process sponsored by the province). Teachers in this school acknowledged the contribution of these resources to their learning through collective reflection on school priorities. Thus, in one case resources were explicitly associated with OL and in another they were not.

Some teachers associated school conditions directly with outcomes, leaving implicit the necessary mediation of their own learning processes. Variation among schools was considerable. Overall, school *culture, structure* and *strategy* were cited with almost the same frequency. *Resources* were cited about half as frequently and *vision* was associated with outcomes in only a few cases.

Among school conditions, in sum, *culture* appears to be the dominant influence on collective learning. School *mission and vision* and *resources* may be less important in fostering OL than commonly is believed to be the case. Both *structure* and *strategy* are associated moderately and about equally with OL, especially collective learning. School conditions as a whole are much more frequently associated with collective than with individual learning.

Organizational learning processes

> This year we were looking at reflection as a group. A teaching partner and myself looked at whether classroom and teacher talk would enhance the reflection and enhance children's growth in the visual arts. We focused on visual arts just because you have to keep fine-tuning it way down to small, small pieces. It was quite interesting. It was something I sort of knew, yes, it would but it's interesting to actually break it down and see how it works and then come to realizations of how much more has to be done and all the rest of it.

Processes through which teachers reported learning distinguishes individual from collective processes. The most frequently mentioned collective process for learning was the exchange of information through informal discussions among colleagues. This occurred when teachers felt comfortable sharing their own learning with others and receiving suggestions for improvement from colleagues. Only in small schools did these exchanges appear to involve whole staffs. More typically, they occurred within smaller groups such as a grade or department teams. Teachers also reported using a trial-and-error approach with new practices, perhaps after jointly participating in a workshop or simply on the initiative of a fellow staff member. Experimentation with new practice was considered to be an effective way to adapt current practice after evaluating what needed to be changed. As important as acknowledging and celebrating "what worked" with such experiments was the problem-solving that arose in response to perceived failures. In most schools teachers also mentioned working with colleagues to develop new curricula or instructional approaches, a more systematic exchange than informal sharing.

Spending time in each other's classes was another means teachers used for their learning, either casually "dropping in" for conversation or through a more formal process of scheduled observation followed by feedback. In only a few cases did teachers with expertise in specific areas demonstrate new strategies for their colleagues or were staff meetings used to practice new methods. Other less frequently used processes were collective reflection on school goals, professional reading, and research. Teachers in most schools visited other

Table 4.1. Characteristics of schools as learning organizations.

School vision and mission
- clear and accessible to most staff
- shared by most staff
- perceived to be meaningful by most staff
- pervasive in conversation and decision-making

School culture
- collaborative
- shared belief in the importance of continuous professional growth
- norms of mutual support
- belief in providing honest, candid feedback to one's colleagues
- informal sharing of ideas and materials
- repect for colleagues' ideas
- support for risk-taking
- encouragement for open discussion of difficulties
- shared celebration of successes
- all students valued regardless of their needs
- commitment to helping students

School structure
- open and inclusive decision-making processes
- distribution of decision-making authority to school committees
- decisions by consensus
- small size of school
- team teaching arrangements
- brief weekly planning meetings
- frequent problem-solving sessions among sub-groups of staff
- regularly scheduled pro-d time in school
- arrangements of physical space to facilitate team teaching
- freedom to test new strategies within teacher's own classroom
- common preparation periods for teachers needing time to work together
- cross-department appointment of teachers

School strategies
- use of a systematic strategy for school goal-setting involving students, parents, and staff (school accreditation was an oft-cited context for this)
- development of school growth plans
- development of individual growth plans which reflect school growth plans
- establishment of a restricted, manageable number of priorities for action
- periodic review and revision of school goals and priorities
- encouragement for observing one another's classroom practices
- well-designed processes for implementing specific program initiatives, including processes to ensure follow-through

Policy and resources
- sufficient resources to support essential pro-d
- using colleagues within one's own school as resources for pro-d
- availability of a professional library and professional readings circulated among staff
- availability of curriculum resources and computer facilities
- access to technical assistance for implementing new practices
- access to community facilities

schools to observe some practice they were about to introduce or were considering for their school.

Individual learning processes reported by teachers focused more on personal reflection, learning from their own personal and professional experience. Teachers also talked about their own experimentation with new strategies and questioning of their assumptions about teaching and learning. Professional reading and library research were used by teachers for professional growth. Reaching out to colleagues in other schools was helpful for some teachers, as was observing what instructional methods were effective for members of their own family—their own children, for example.

On average, teachers made more than three associations between OL and one or more outcomes; collective and individual learning were equally associated with such outcomes, and variation among schools was significant. Teachers in two schools more frequently associated outcomes with collective learning than with individual learning; this reflects talk about constant exchange of ideas in a culture of continuous growth in the first school and the availability of opportunities for joint learning in the second. At the same time, however, evidence from another school demonstrated the highest incidence of OL not associated with outcomes for various reasons (e.g. unwillingness of colleagues to implement new strategies, ineffectiveness of a learning activity, lack of opportunities to acquire appropriate knowledge).

By way of summary, teachers learned through their informal, daily contacts with other teachers and through reflecting on their own classroom experiments. Organized, formal professional development time also was quite important for them, however, when it directly addressed their own felt needs and when there were opportunities to "socially process" the professional development experience (share it with others in some fashion).

Outcomes of organizational learning

> I've gone from very individual and almost lecture-oriented teaching to very student-directed and cooperative. That kind of thing. That's been a huge change in my teaching. That's been an evolution of a few years.

Most of the outcomes of OL reported by teachers were *practices*, followed by *understanding* and *commitment*, with *new skills* a distant fourth. The discussion of new skills was mostly about improved teaching techniques, the precise nature of which was not specified. As well, a few teachers said they had acquired better management skills that led to a smoother day-to-day flow in their classrooms. Some teachers said they had acquired new computer skills for use with their classes or had improved their techniques for working with special needs students.

The *new understanding* mentioned most frequently was acceptance of the necessity of meeting the needs of each individual student and the importance of relating to the "whole" child and not only his/her academic development. Teachers also gained a new awareness of which instructional practices were effective and which were not, perhaps a result of their reflection on current practice as a process for learning. Greater familiarity with a variety of instructional approaches and of reasons for changing approaches were other outcomes, as was an understanding of how learning varies for different students.

Less than 10% of teachers reported other outcomes, such as: gaining an understanding of how to relate to immigrant students who came without English-language skills; what made a new program or teaching strategy better for students; how goals helped to focus work; and what influences student learning. Overall, knowledge gained through OL was spread between a broader perspective on technical aspects of teaching and a better understanding of students.

Increased commitment was a third category of outcomes evident in statements reflecting teachers' pleasure that they were enhancing student learning by making it more exciting and authentic. Teachers also showed evidence of professional commitment in their desire to continue professional growth and to do the best work possible. Teachers talked about their preference for new practices or programs over what they had been doing previously and about the excitement of trying new approaches compared with repeating past practice. Some teachers said they had been empowered as professionals with new expertise or by recognition from others. In four schools, individual teachers talked about their commitment to their school and the pleasure they received from being part of that school. Several teachers said they were excited by their school's vision or goals and others talked about their enjoyment of teaching in general or the excitement of exposure to a new idea.

Most teachers provided at least one example of a *new practice* they had implemented over the last few years. Fifty percent of the teachers indicated they were implementing new practices or updating their practice, although the specific nature of the new practice was not defined. When teachers talked about specific changes, the change most likely to be described was increased use of cooperative learning strategies with their students, an innovation mentioned by one third of the teachers representing all six schools. A general move to a child-centered approach was reflected in examples from all six schools of more active involvement of students in their learning through giving them more choice in the content of their curricula and by redefining the teacher role to be more that of a facilitator.

A more deliberate attempt to address the needs of the whole child was another outcome, as was the attempt to meet individual student needs by individualizing programs, including special needs students, or changing practice to accommodate a new type of student such as was the case in schools with an increase in ESL students. Some teachers also talked about being more flexible in their practice and more open-ended in their approach to curriculum. More project or group work was used to provide a variety of learning experiences. New evaluation strategies such as greater use of anecdotal reports, student-led conferences, or self-evaluation were adopted for compatibility with new programs.

Of the OL outcomes identified by teachers, in sum, the category *new skills* was mentioned least frequently. When mentioned, it usually meant new instructional strategies. *Understandings* arising through OL primarily were of two sorts: a broader perspective on instructional techniques and a better understanding of students. Increased commitments were reported by teachers: to student learning, to their own professional growth, and to their schools. New practices reflected a move toward more child-centered and flexible instruction, new forms of student assessment, and other practices closely akin to Ministry policy directions.

Leadership practices which foster organizational learning

Our perspective on questions about leadership was influenced by the expectation that transformational forms of leadership might be especially helpful in fostering both individual and organizational learning in schools. With its roots in non-school organizations (e.g., Bass, 1985; Burns, 1978), this form of leadership aims to significantly enhance both the commitments and capacities of those who experience it. The small body of evidence recently collected in school contexts has suggested promising effects when schools are immersed in major change projects such as the restructuring efforts of the six schools in this study (Leithwood, 1994). In the remainder of this section, we identify eight specific dimensions of such leadership and describe the contribution teachers believed each made to their individual and collective learning.

Identifies and articulates a vision

This dimension encompasses practices on the part of the leader aimed at identifying new opportunities for his or her school and developing (often collaboratively), articulating, and inspiring others with a vision of the future. Of the fourteen associations teachers made with the principals' vision building, nine were associated with school conditions. Three teachers in one school and one in another talked about how the principal's vision influenced the school vision. Teachers in three schools mentioned such influence on school strategy development. One reference was made to school culture. The following comment captures teachers' views of how the philosophy of inclusion of the principal in one school influenced programs and subsequently school structure:

> This school has had a huge push in the last few years, driven very much by the philosophy of our principal, to accommodate individual differences in learning to a huge degree. What that means is making a branch of our school accommodate the learning disabled. We have a learning assistance area and a learning for the disabled area that has just exploded in size.

This statement from a teacher in one school highlights the powerful impact a principal's vision can have on the culture of the school (a relatively rare occurrence):

When this administrator first joined us, we came up with our mission statement, if you like. That has very much formed the culture. I think that colleagues in other schools know about that culture and know about the kind of almost charisma that exists at this school. When we have student teachers come they are told by their advisors that this kind of experience at this school is one they may never experience again.

Several teachers described how the principal's vision directly influenced collective learning. For example:

I think that she probably has a clear vision of what she wants to do for the school and where she wants to go. From her comes a little bit of direction . . . I think that from her vision comes suggestions and changes and ideas for maybe how we want to change ourselves. For example, multi-ageing and cooperative learning and all the things that have been brought in in the last few years have been by and large from [our principal's] vision of what a primary school should be like.

Fosters the acceptance of group goals

This leadership dimension includes practices on the part of the leader aimed at promoting cooperation among staff and assisting them to work together toward common goals. Although there was at least one teacher comment from every school affirming their principal's role in goal development, most of the comments simply indicated that the principal initiated the process, was a member of the goal-setting committee, or asked for input. Only one principal was reported to be actively involved in developing goal consensus. One of the teachers in that school said, "we all seem to want similar things; we're kind of working towards the same goals." That same principal was also the only one who was viewed, by three teachers, as helping staff develop individual growth plans. During goal-setting, the principal in one school fostered OL by encouraging staff to systematically reflect on the activities of the past year. One teacher clearly articulated the process:

We certainly start by celebrating the year; that was on that warm June day when we get together and find out what was good about that. We do that starting in small groups and then meeting in larger groups and ever larger until all the common things are weighed out. We see from those things that worked for us if there are any that we still want to work on and continue into the next year. Then we problem solve some of the areas where we feel that we need to grow and where there has been problems in the last year. We work through a process so that we can come up with the ones that have the most weight for us and are of the most importance. Those become our growth plan for the next year.
Q: Quite organized?

R: Definitely. Yes.
Q: What's the principal's role?
R: She's the facilitator. She usually sets up the process for us.

Conveys high performance expectations

Included as part of this leadership dimension are practices that demonstrate the leader's expectations for excellence, quality, and/or high performance on the part of staff. The interviews provided relatively little evidence that principals in the six schools, as a group, held high performance expectations for their staffs, at least that their staffs could detect. Only nine teachers in three different schools made comments about such high expectations. These principals, it was reported, demanded high "professionalism," held high expectations for professional growth, and were committed to keeping the school on the cutting edge of changes in education. Encouraging teachers to be creative and to try new strategies were indicators of high expectations to teachers.

Several specific practices were perceived to influence both school conditions and OL. One teacher said that her principal's commitment to fulfilling provincial mandates, and keeping the school on the cutting edge of changes in education, encouraged the staff's commitment to the same vision. Another teacher in the same school said that her principal's expectation that staff will try new teaching strategies (along with other factors) influenced her to learn about them. In another school, professional growth was expected. The following excerpt shows how high expectation (along with the stimulation of reading research articles) encouraged learning and subsequently changed practice.

> Our teaching practices change all the time because we end up sitting down and reflecting so much. You're constantly reviewing and learning. Our administrator throws out enough research articles and this and that and calls upon us to lead staff meetings. We're just expected to keep up with what's happening and really talk about it. You're constantly changing and reassessing everything you do.

Provides appropriate models

This dimension of leadership encompasses practices that set examples for staff to follow and which are consistent with the values espoused by the leader. Many teachers in five out of six schools believed that their principal was a good role model. These principals set an example by working hard, having lots of energy, being genuine in their beliefs, modeling openness, having good people skills, and by showing evidence of learning by growing and changing themselves. Three principals were perceived by at least one teacher in their schools as symbolizing success and accomplishment in the field; for example, by getting grants or by being well known in the field of global education. Being involved in all aspects of the school and showing respect for and interest in the students also was considered to exemplify the modeling of excellence. One principal modeled good instructional strategies in the classroom.

The principal in one school helped her staff learn by modeling teaching strategies in the classroom and in staff meetings. A different teacher in the same school described how, by modeling overall excellence, the principal contributed to organizational learning.

> [The principal stimulates professional learning] in such a gentle way. Never does she ask anybody to do anything. She sort of throws questions at you. She's asked us to give little workshops within our own school, which I have done. She always sends you these little notes on things that you wouldn't even have thought that was part of what you have done above and beyond the call of duty that she's spotted. What you end up doing is, when you're reading a book, like *Revolution from Within*, I would find little things that connected up with education and I'd xerox them and put them on her desk. There's always a dialogue or something going on. Somebody is feeding you back all the time. She has such an excellent sense of humour that she's just a comfortable human being to be with. She sees the light side of things. Plus, she is so into what children are all about and really looks behind the scenes and everything. It seems so ridiculous to go on about it, but she's the first administrator that I totally admire and enjoy. I mean, I'd do anything just because she asked because I know that it's going to be worthwhile and it's going to help me grow as a person. She obviously learns from all of us too. It's a real give and take situation.

This same teacher also described how the principal's gentle manner of drawing people out led to her commitment to do the same thing in her classroom. One teacher in another school commented that having the principal attend workshops in the summer motivated teachers to attend as well. This was seen to be a contribution to a culture of learning.

Provides individualized support

This is a dimension of leadership represented in practices on the part of the leader that indicates respect for individual members of staff and concern about their personal feelings and needs. Most teachers in five of the six schools indicated that their principal provided support for professional learning. Typically, this meant providing resources for learning in the form of money, books, furniture, or materials. Teachers in several schools considered their principal to be particularly adept at procuring funds. Said one teacher, "The principal is a magician in terms of finding money and grants." Other teachers reported that their principals even used their own administrative professional development funds for things that teachers needed. Other kinds of tangible support for professional learning included providing release time or other scheduling help, sharing information or finding speakers, and encouraging participation in decision-making by collecting and distributing information: "He does the groundwork for us," claimed one teacher.

Providing moral support was mentioned by many teachers from five schools. There is the sense that these principals "are always there for us," do whatever they can to get staff what they need, and generally support what teachers do. One teacher said, "we all feel that whatever we do she's going to be supportive of it as long as it's not harmful to kids." Sometimes this support was shown by an eagerness to listen, being accessible, fair, open, and sympathetic. Sometimes support was shown by offering positive reinforcement which made staff feel appreciated. Support was also shown in the form of encouragement to try, to take risks. And leaders' signs of appreciation were reported to build a collaborative culture:

> Administration is very good at thanking people for their participation in terms of sending thankyou notes and taking the staff out for a drink or something. All of that builds a sense of collegiality in the staff.

Provides intellectual stimulation

This dimension of leadership includes practices that challenge staff to re-examine some of the assumptions about their work and to rethink how it can be performed.

About a quarter of all teachers interviewed claimed that their principals provided some sort of intellectual stimulation. This stimulation was the kind that made them re-examine their practices:

> She certainly encourages us to think about what we do. I must say, with [our principal] we can't just go about hiding in our classrooms doing our own thing. [She] forces us to take a good look at ourselves.

Intellectual stimulation also meant passing on information from journals or other sources, bringing new ideas into the school, and providing professional development at staff meetings. As one teacher explained:

> That's helpful to realize that a lot of what we were basing our teaching on has been proven incorrect. Children don't need to be totally quiet or structured in roles and things like that in order to learn. There's many styles and ways of learning and teaching and different ways of representing knowledge.

Other forms of intellectual stimulation included organizing and chairing professional development sessions, finding out what staff needed to learn, encouraging staff to put on workshops or to lead staff meetings, and discussing a teacher's progress with personal growth goals.

Builds a productive school culture

This category of practices encompasses behavior on the part of the leader that encourages collaboration among staff and assists in creating a widely shared set of norms, values, and beliefs consistent with continuous improvement of services for students. Many teachers did not consider their principals to have much influence on school culture. But four principals were perceived by at least some of their teaching colleagues as being fundamental to that culture. Inspiring respect, being kind, thoughtful, sincere, honest, and hard working were attributes that contributed to this perception. One teacher observed:

> I think that she's the main factor in the fact that we've all been quite successful in how we've changed.

Demonstrating an interest in the students and clearly setting their needs as a priority was considered to be an important influence in three schools. For example:

> He's terrific about coming around to the classroom, looking at the children's work, and it's never negative things; it's always positive things. And he feels very comfortable, and we feel comfortable, about bringing children to show him special things that they have done.

A strong belief in the value of communication and collegiality and a willingness and ability to be flexible were considered to be characteristics conducive to a collaborative culture. Teachers also valued principals who showed them respect, treated them as professionals, and who were an integral part of the staff. Being seen as working more for the school than for the school district was mentioned by one teacher as being important. Hiring staff who share the same philosophy or who can work well with existing staff was mentioned by three teachers as a way that their principals contributed to a collaborative culture. Other ways principals influenced culture was by encouraging parental involvement, fostering professional development, and being accepting of different cultures.

One teacher claimed that, although the principal doesn't get involved very much with school culture, that affects structure in a positive way because "a lot of the influence is out of his hands and in the hands of people who are really working in the school."

The striking impact of a principal's close involvement with staff is evident in this remark from a teacher:

> I think that [the principal is] the main factor in the fact that we've all been quite successful in how we've changed. She's been there when we've been in tears and she's been there when we were happy and ecstatic about something that happens and she's there to kind of celebrate with us.

Teachers in two schools credited their principals' philosophy with fostering their child-centered culture. One teacher said her principal sets that tone by "putting student needs above timetable needs," for example. Another said that respect among staff to students is engendered by the administrator's actions. A third teacher in that school said the principal's actions demonstrated clearly the importance of being understanding of people from many nationalities and backgrounds. In one school, the principal's strong belief in collaborative decision-making fostered a collegial culture:

> I think it helps teachers to feel like we're all participants together in what happens and what's emphasized and the learning that takes place. I think it fosters a healthy environment.

Helps structure the school to enhance participation in decisions

Included in this dimension of leadersip are practices which create opportunities for all "stakeholder" groups to participate effectively in school decision-making. More teachers talked about leadership influences on school structure than on any other school condition. The following remarks from teachers in two different schools exemplify how this influence was exercised:

> The principal is always asking for input and generally takes it into consideration if it's a feasible kind of situation.

> I don't feel there are very many decisions made about the school over which I don't have some input. There's opportunity for even more if I wanted. It's up to me. My ideas are always sought.

In spite of the high degree of collaboration exhibited in these schools, several teachers noted that their principals could make unilateral decisions when appropriate or for efficiency. Principals were seen to encourage participation on committees and to support the committee structure by being actively involved and by organizing or spearheading activities. Many teachers applauded the autonomy their principals gave them to make their own decisions in certain areas. Three principals shared power and responsibility by asking teachers to give workshops, lead staff meetings, manage the budget, and by delegating many duties to the vice-principals.

To facilitate teacher collaboration, six teachers in two schools said their principals altered working conditions by making changes to the physical plant (e.g. creating convenient meeting rooms), restructuring the timetable (e.g. creating X block), and by arranging for leadership positions specifically designed to foster their learning.

In sum, evidence from the study indicates that all dimensions of transformational leadership contribute significantly to school conditions fostering OL as well as to OL directly. There were almost twice as many associations made by teachers, in the interviews, between leadership and school conditions as between leadership and OL. This lends support to the assertion that the effects of leadership are often indirect.

Summary and conclusions

While organizational learning is a perspective frequently used to better understand non-school organizations, it has rarely been applied to schools. As a consequence, we have almost no systematic evidence describing the conditions which foster and inhibit such learning in schools. The main purpose of this study was to begin to address that gap in evidence. This was done largely through interview data from 72 teachers and 6 principals in six schools. Results of the study give rise to five broad assertions with which we conclude.

School leadership practices have among the strongest direct and indirect influences on OL

No other sets of conditions outside the school, for example, were associated by teachers with the creation of school conditions fostering OL as often as school leadership. And leadership was identified as a direct influence on OL about as frequently as school structure and only less frequently than school culture. More than any other set of conditions, however, teachers identified some leadership practices as impediments to OL. Those practices associated with a transformational model of school leadership made uniformly positive contributions to OL.

Organizational learning processes are highly varied

Evidence suggests that teachers learn through quite informal means from their colleagues and their own individual classroom experiences. As well, however, they learn through such formal structures as scheduled professional development inside and outside the school and visits to other schools. Organizational learning, then, seems to be fostered by a rich menu of opportunities of both a formal and informal nature. Such opportunities will be helpful to the extent that they address problems of acknowledged concern to teachers and under conditions in which there is an opportunity for teachers to socially process the information they encounter.

These conditions for teacher learning, although hardly surprising, are not well reflected in current practice. Professional development opportunities for teachers often focus on issues or problems identified by people other than teachers and frequently occur in ways that leave teachers isolated in their subsequent efforts to make sense of what they learn for purposes of their own practice.

District contributions to organizational learning in schools are underestimated

Evidence from the larger study indicates that school principals are significantly influenced by district-level decision-making. But beyond this, our data show what, for some, may be a surprising amount of influence exercised by districts on the organizational learning of teachers themselves. This, of course, was not the case in all schools. What seems to be critical in order for districts to have such influence are their professional development policies and resources. Districts provide opportunities for teachers through the in-service they make available. That is an important first step in their influence on teachers.

When districts also have policies which help ensure the social processing of new ideas, this further contributes to districts' influence on organizational learning in schools. For example, two of the schools in our study belonged to districts in which attendance at district workshops by pairs of teachers or teams of teachers from individual schools was encouraged. Teachers viewed that practice as a significant element in their making sense of the new ideas presented to them during such professional development.

A coherent sense of direction for the school is crucial in fostering organizational learning

This study provided several examples of schools in which teachers believed they had a relatively clear understanding of the general direction in which the schools were headed, and other examples in which there was little such sense of direction. Schools with a coherent sense of direction eventually were able to make sense of even relatively large numbers of disparate initiatives undertaken within the school, as is often the case in larger schools. But equally large schools without a coherent sense of direction eventually appeared to be "spinning their wheels," in spite of many creative initiatives. The amount of organizational learning that took place in these schools seemed minimal, even though individual teacher learning might have been significant.

It seems possible, then, for a school to operate much of the time in a relatively balkanized manner and still make progress in terms of its organizational learning providing staff members share some sense of overall purpose for the school.

Sources of a school's coherent sense of direction are not obvious

Neither the vision-building activities of school leaders nor the school's mission and vision, as it was defined in this study, appeared to stimulate significant OL. Even the apparently most effective school leaders we studied, for example, were not identified as spending much time articulating or building an explicit school mission or vision. But there were other less obvious sources of a coherent sense of direction in this school, including goal-setting strategies that the leader often initiated focused on shorter-term directions for the school. Other sources of direction included the school culture and a coherent set of practices engaged in by leaders which modeled at least the leader's vision of what the school should become.

This study provides a quite detailed picture of the causes and consequences of organizational learning in schools. It has identified the specific processes that teachers use for their own collective and individual learning. And this is an important first step in a program of research aimed at testing the power of an organizational learning lens to explain variation in school restructuring success. Of course, evidence from six schools hardly provides the "final word" on the causes and consequences of OL in schools, so comparable efforts in many more and varied schools is an important step for future research. However, the "shape" that OL takes in schools may now be sufficiently visible to initiate, in parallel, the fundamental task of assessing the effects of OL on student growth. Do variations in the processes and outcomes of organizational learning explain significant variation in student effects? The case that it ought to is theoretically compelling: as yet, there is no empirical evidence that it does.

References

Argyris, C. & Schön, D.A. (1978). *Organizational learning: A theory of action perspective*. Reading, MA: Addison-Wesley.

Bass, B. (1985). *Leadership and performance beyond expectations*. New York: The Free Press.

British Columbia Ministry of Education (1989). *Year 2000: A framework for learning*. Victoria, B.C.: Queen's Printer for British Columbia.

Burns, J. (1978). *Leadership*. New York: Harper & Row.

Dixon, R.E. (1992). *Future schools and how to get there from here: A primer for evolutionaries*. Toronto: ECW Press.

Cousins, J.B. (in press). Understanding organizational learning for school leadership and educational reform. In K. Leithwood (ed.), *International handbook of educational leadership and administration*. Netherlands: Kluwer Academic Press.

Fiol, C.M. & Lyles, M.A. (1985). Organizational learning. *Academy of Management Review* 10: 803–813.

Hedberg, B. (1981). How organizations learn and unlearn. In P.C. Nystrom & W.H. Starbuck (eds.), *Handbook of organizational design, volume 1: Adapting organizations to their environments*. New York: Oxford University Press.

Leithwood, K. (1992). The move toward transformational leadership. *Educational Leadership* 49 (5): 8–12.

Leithwood, K. (1994). Leadership for school restructuring. *Educational Administration Quarterly* 30 (4): 498–518.

Leithwood, K. & Aitken, R. (1995). *Making schools smarter: A system for monitoring school and district progress*. Newbury Park, CA: Corwin.

Leithwood, K., Dart, B., Jantzi, D. & Steinbach, R. (1990). *Implementing the primary program: The first year*. Victoria, B.C.: Final report for Year One submitted to the B.C. Ministry of Education.

Leithwood, K., Dart, B., Jantzi, D. & Steinbach, R. (1991). *Building commitment for change: A focus on school leadership*. Victoria, B.C.: Final report for Year Two submitted to the B.C. Ministry of Education.

Leithwood, K., Dart, B., Jantzi, D. & Steinbach, R. (1993a). *Fostering organizational learning: A study in British Columbia's intermediate developmental sites, 1990–1992*. Victoria, B.C.: Final report for Year Three submitted to the B.C. Ministry of Education.

Leithwood, K., Dart, B., Jantzi, D. & Steinbach, R. (1993b). *Building commitment for change and fostering organizational learning*. Victoria, B.C.: Final report for Year Four submitted to the B.C. Ministry of Education.

Leithwood, K., Dart, B., Jantzi, D. & Steinbach, R. (1994). *The development of schools as learning organizations*. Victoria, B.C.: Final report for Year Five submitted to the B.C. Ministry of Education.

Levitt, B. & March, J.G. (1988). Organizational learning. *Annual Review of Sociology* 14: 319–340.

Naisbitt, J. & Aburdene, P. (1985). *Reinventing the corporation*. New York: Warner Books.

Perkins, D. (1992). *Smart schools*. New York: Free Press.

Rosenholtz, S. (1989). *Teachers' workplace*. New York: Longman, Inc.

Schein, E.H. (1990). Organizational culture. *American Psychologist* 45 (2): 109–119.

Schlecty, P.C. (1990). *Schools for the 21st century*. San Francisco: Jossey-Bass.

Senge, P. (1990). *The fifth discipline: The art and practice of organizational learning.* New York: Doubleday.

Watkins, K.E. & Marsick, V.J. (1993). *Sculpting the learning organization.* San Francisco: Jossey-Bass.

Part II

Interventions to Stimulate Organizational Learning in Schools

5

Talking about Restructuring: Using Concept Maps

Jean A. King, Jeffrey Allen, and Khahm Nguyen

Origin of this mapping effort in the context of organizational learning

Organizational learning has surely earned a place on the top ten list of concepts in the current reform literature, and its widespread popularity among scholars points to its appeal. If an organization can process and act on information in a meaningful way, the theory goes, its functioning over time should improve and its vision become more attainable. Organizations that are quick studies will adapt more rapidly than those that are learning disabled, making specific techniques and strategies for organizational learning important means to the overall end of continuing improvement. Concept mapping, a graphic technique for understanding people's thoughts and how they are linked with each other (Hart & Bredeson, 1995; Khattri & Miles, 1993; Margulies, 1995; Oldfeather, 1994), is one such technique. This chapter presents the results of a two-year project that sought to increase organizational learning in an urban school, Thomas Paine High School (a pseudonym), through the use of such mapping.

It is one thing to talk about concept mapping as a means of increasing organizational learning; it is another thing altogether to bring it to life in a school, even in a school where a thoughtful faculty has enthusiastically participated in change efforts for a number of years. Philosophical rationales and well-constructed arguments sometimes fall apart in practice, and at an urban school day-to-day immediacies can easily overwhelm longer-term discussions of change, especially if they are not perceived as useful. The concept mapping effort discussed here consisted of two phases.

1)   an initial, pilot effort, involving three teacher leaders, that asked the question: To what extent is concept mapping a viable tool for fostering discussion of change issues in an urban school?
2)   an expanded effort, involving other staff and administrators, that addressed the question: To what extent can the process of concept mapping be integrated into ongoing discussions of change in an urban school?

While the results were somewhat different from our expectations, they may nevertheless be instructive for those seeking to use this technique in school settings.

A passing review of related literature

Concept maps are concerned with relationships among ideas. "In concept mapping, ideas are represented in the form of a picture. To construct the map, ideas first have to be described or generated, and the interrelationships between them articulated" (Trochim, 1989, p. 1). The visible grouping of these relationships helps to reduce the complexity of a set of concepts, aiding people in the acquisition of knowledge, the analysis of situations, and effective problem-solving (Posner & Rudnitsky, 1986, p. 68). The literature describing the use of concept mapping is of two types: that related to student learning and curriculum development; and that related to program evaluation and planning.

Student learning and curriculum development

The use of maps emerged out of a debate in science education that focused on whether or not children could fully understand abstract concepts (e.g. matter, infinity, energy). Mapping is based on a theory of meaningful learning. It "relates directly to such theoretical principles as prior knowledge, subsumption, progressive differentiation, cognitive bridging, and integrative reconciliation" (Wandersee, 1990, p. 927). Curriculum developers have created concept maps for textbooks and teachers' guides, and teachers have used these in classrooms as an instructional technique to help students see holistically the relationships among concepts. Maps have been used in teaching earth science, mathematics, physics, statistics, and biology (Barenholz & Tamir, 1992, p. 38), as well as in teacher education (Morine-Dershimer, 1991, cited in Khattri & Miles, 1994). "Concept mapping has become an important tool to help students learn to learn meaningfully, and to help teachers become more effective teachers" (Novak, 1990, p. 941).

Program evaluation and planning

Trochim (1989) describes how concept mapping—a process of "structured conceptualization"—can serve to articulate program theory in context. In a highly technical process guided by a facilitator, a group responsible for a given evaluation and planning effort develops a collaborative map and ultimately decides how it will be used. Trochim's mapping process involves six steps:

1)   selecting participants for the group process
2)   initial brainstorming of concepts
3)   structuring of the statements using an unstructured card-sorting proce-
      dure, the results of which are combined across people
4)   multidimensional scaling and cluster analysis, leading to a "point" or
      "statement map," a "cluster map," a "point-rating map," and a "cluster-
      rating map"
5)   interpretation of the maps
6)   discussion of how to use the final summary map

The process that creates the map makes visible a unified theory of action for the program, which involves "figuring out what are all the things that have to happen in the program, in what order and sequence, to achieve various outcomes and impacts, in what order and sequence" (Patton, 1989, p. 377).

The project described in this paper used Khattri and Miles' (1994) application of concept mapping to school restructuring, which combines the two general approaches found in the mapping literature. It sought to assist those involved in restructuring efforts to learn both what key concepts were guiding their effort and how these were related. It also served a program evaluation and planning function in that different individuals were able to compare their personal maps to determine commonalities and differences.

The school context

Thomas Paine is an urban high school with a current enrollment of around 1,000 students, grades 9–12, in a major metropolitan area in the Midwest. It serves two neighborhood communities: one—its traditional student population—a blue-collar, white community; the other a more recent and less stable African-American community. In addition, many middle-class students attend an academically challenging International Baccalaureate (IB) magnet program housed in the building. The continuing presence of a district-wide program for severely emotionally and behaviorally disordered (EBD) students at Paine insures a small, but highly visible population of troubled students.

Until a recent renovation, peeling paint was the single most noticeable feature of the school, but this detail inaccurately suggested that Thomas Paine lacks resources. By comparison to schools in other large cities, its facilities are impressive. For example, teachers in most subject areas at Paine have adequate numbers of texts and instructional materials for their students (although they are quick to note that more are needed!), and the school has a well-staffed and active media center. Many Paine classrooms are sunny and large, and the halls are decorated with student work, newspaper articles, and brightly-colored posters and art. The school's gymnasium and athletic fields, along with access to a nearby ice arena, support its physical education programs and both male and female team sports.

Although recent indicators point to improvement, for much of the past decade Paine ranked at the bottom of the district in academics. With unfortunate consistency, Paine's scores were at the bottom of positive measures (e.g. achievement test scores) and at the top of negative measures (e.g. absences, student turnover). Pressure from outside gang-related activities—or at least the threat of gang violence—remains an issue in the school, and aides with walkie-talkies blaring routinely patrol the halls. A uniformed officer, often present in the halls, serves as liaison to the police department. Unlike counterparts in other cities, however, Paine is not a dangerous school. Weapons are not commonplace, and faculty and students alike were horrified two years ago when a gang member was caught near the building with a loaded and cocked gun.

Paine has a teaching and support staff of approximately 45, almost entirely white. In 1994–1995, the student population increased dramatically, from 730 students to nearly 900, with a racial/ethnic mix of almost half African-American, a slightly smaller percentage of whites, and the remainder a mixture of other minorities (Asian-American, American-Indian, and Hispanic). Staff attributed the increase in part to the growth of the IB program, but also to increased student retention. The student population is now fairly stable at 1,000.

During the two years of the concept mapping project Paine was led by a four-person administrative team: Principal James Hissop, a Native American; two African-American male APs, one of whom left mid-project and was replaced by a white male; and a third AP who sometimes labels herself the "token white female." The school enjoys two major collaborations. First, its active business partner is a local bank that sponsors a number of activities (e.g. the Renaissance student recognition program, a mentorship program, teller training, and scholarship support for graduates). Second, over the past five years Paine has forged a professional development school (PDS)—the only secondary PDS in the state—with a local College of Education.

Paine's school day begins at 7:15 a.m., requiring most teachers to get up well before 6:00 a.m. in order to arrive at school with enough time to organize for the day; during the winter, everyone arrives in the dark. "Lunch" begins at 10:39 a.m. Because of a high level of student absenteeism (15 to 20% of the student body continues to be absent any given day), actual class sizes tend to be small on any given day, but irregular attendance can make teaching difficult. In addition, Paine continues to experience a fairly high rate of student turnover so that class rosters throughout the year are in steady flux as student names are added and deleted.

Not surprisingly, when the school day officially ends around 2:00 p. m., Paine's teachers are often exhausted, but faculty are committed to attending after-school meetings of a variety of types—whole school staff meetings, committee meetings (e.g. staff development, Outcomes Accreditation, Site Council), team meetings, and departmental meetings, in addition to meetings with individual students, teachers, and parents. As one teacher noted, Paine is no longer a "place [for teachers] to hide out," and those uninterested in change are opting to retire or to leave the school. When the concept mapping intervention began after five years of restructuring activity, the sense of progress—and the growing impatience with the fact that a great deal remained to be done—was palpable.

The mapping intervention at Paine: Phase One

In the context of Paine's professional development school, activities over the course of five years pointed to certain teacher leaders as likely first cases for creating maps. We began with three key leaders, each of whom had taught at the school for over a decade and had been an active participant in the restructuring process from its inception. In initial discussions of the mapping effort, the principal identified these three women as important sources of information, and the ongoing experience at Paine of one of the authors supported this choice. One was a social studies teacher, the other two were math teachers; all had reputations as excellent teachers. Each played a leadership role in the school: one was the school-side co-coordinator of the PDS; one headed the school's Renaissance Program, a recognition program supported by Paine's business partner; and one served as implementation coordinator for an innovative curriculum involving Paine's math teachers. More importantly, these teachers interacted with their colleagues on a regular basis regarding professional issues; they served as mentors and peer coaches. One also ran a collaborative, year-long seminar for the building's teacher residents.

The initial step in the mapping process consisted of three one-on-one interviews with each teacher (lasting from 30 to 65 minutes for the first, 10 to 15 minutes for the second, and even less time for the last). Adapting the mapping process of Khattri and Miles (1994), we focused on Paine's goals and the process for achieving them, in an effort to probe issues central to change in the school. The audiotapes from the first interviews were transcribed, and from them we identified the phrases and issues of importance to that teacher's notion of Paine's goals and a desirable change process. Each phrase was then glued on a post-it note, with the expectation that during the second interview the teachers would create their own maps by arranging the post-it notes on poster board in an appropriate graphic arrangement.

Our initial effort with this process, however, was extremely time-consuming; it was not easy for the first teacher to group her ideas, place her words on poster board, and then label the concepts, especially in the time she had available. To save time in crowded schedules, we altered Khattri and Miles' process by developing a draft map for the remaining two teachers *before* the second interview. Because the interviewer was a regular fixture of school life who had collaborated with these teachers for several years, we were able to build on a high level of trust and an extensive shared knowledge base. The second interview then became a revision of the draft, supplemented by reviewing quotations from the transcript, the teacher's explication of the text, and editing of the proposed map (adding, deleting, and rearranging ideas). The third discussion was a straightforward examination of the revised map to confirm that we'd "got it right"; in one case the teacher did this independently, editing the map and returning it to our mailbox in the main office. All three teachers reportedly enjoyed the process, one noting that it really helped to zero-in on the reasons she was feeling discouraged that year with the continuing change efforts in the building.

Three triggering maps

Figures 5.1 through 5.6 present the individual concept maps created from the teacher leader interviews. The first teacher's two maps document the thinking of a career professional who has been extremely active in restructuring Thomas Paine. Her notion of school goals (Figure 5.1) holds an implicit process. First, get students to attend school, and, if they don't, provide them with alternative ways of doing so. Next, create a climate conducive to learning in the building by making sure it is safe (always a concern at Paine), that people in it have a sense of community, and that all staff have explicit and high expectations. In this map, goals—both social/affective and academic/cognitive—come from a variety of sources (the community, the district, the state, etc.). The final box is student achievement, measured in a number of ways (the NCA Outcomes Accreditation process, grades, and other forms of assessment).

This teacher's process map (Figure 5.2) distinguishes clearly between right and wrong ways to do things. The wrong ways include ineffective meetings, decisions poorly made (i.e., simply being made or made with limited input), the frenetic school context that includes administrative turnover, and no unifying sense of vision. Her sense of how to go about change includes a clear vision plus an overriding process—with timeline—for bringing it to life. She has specific processes in mind: a unified administrative team that communicates effectively as it restructures the entire school; staff and student development during the school year, including mentoring of teachers; topic-based faculty forums for the discussion of ideas prior to voting; summer pay for those teachers willing to develop plans for the year; and, ideally, a dual leadership scheme, with one curricular leader to focus on teacher and student learning and one administrative leader to set building policy.

The second set of maps emphasizes different ideas. The goals map (Figure 5.3) has three sections. A worst-case piece traces potential negative actions by Paine students: they're not in school at all; they're in school, but not in class; they're not behaving, either in class (hence not learning themselves and simultaneously disrupting others) or out of class (not completing assignments, too little parental involvement). The second piece represents the ideal case for students, where ultimate goals—life skills—and high expectations form the basis for a process of education, i.e., students attending school, behaving both in and out of class, supported by their parents (and implicitly by their teachers), culminating in their achievement. At the same time, this teacher highlighted another goal: school improvement, especially relating to Paine's image. Specific ideas for improvement included the physical plant (eliminating Paine's famous peeling paint), keeping

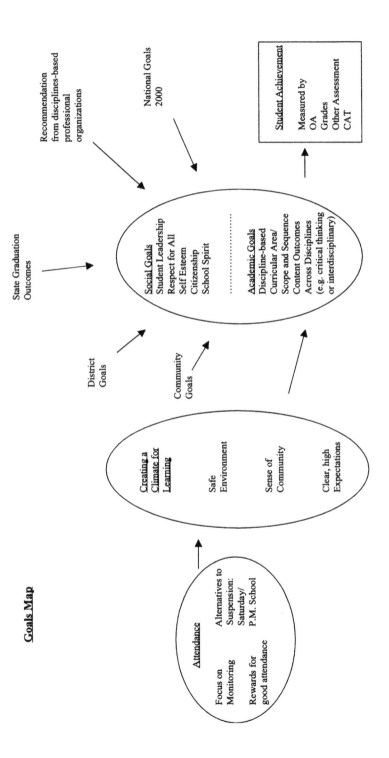

Figure 5.1. Goals and concept map for one teacher.

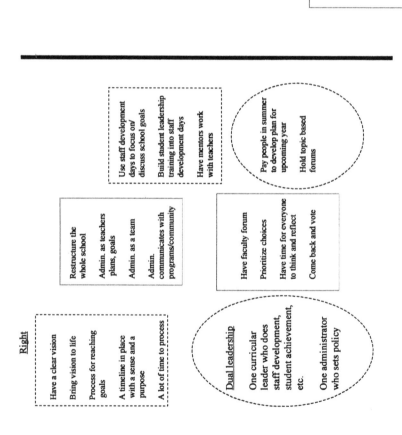

Figure 5.2.   Process concept map for one teacher.

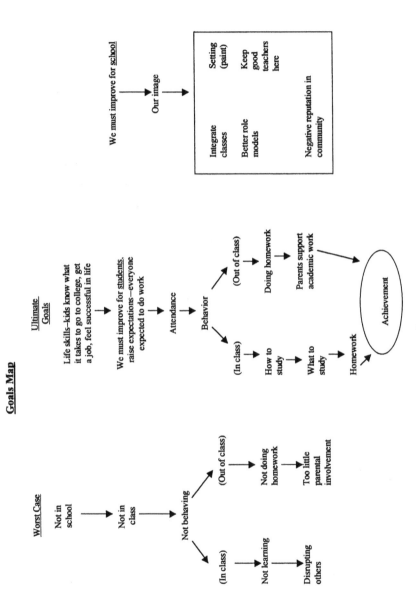

Figure 5.3.   Goals and concept map for second teacher.

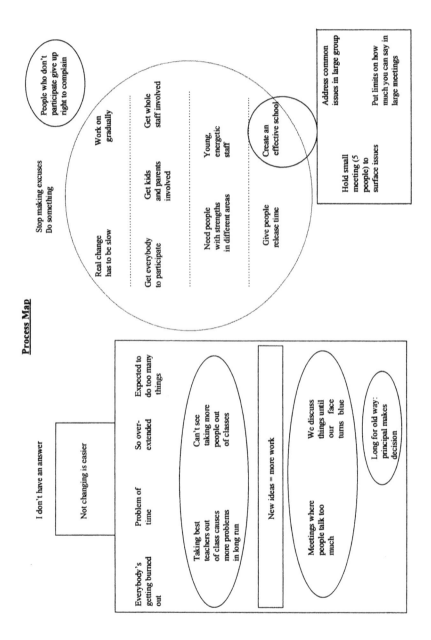

Figure 5.4.   Process concept map for second teacher.

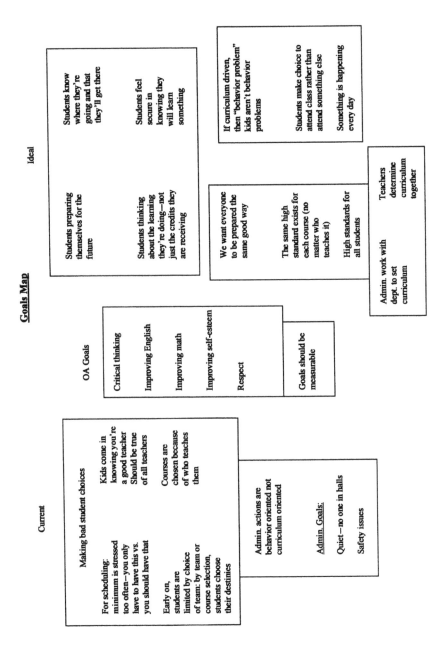

Figure 5.5.   Goals and concept map for third teacher.

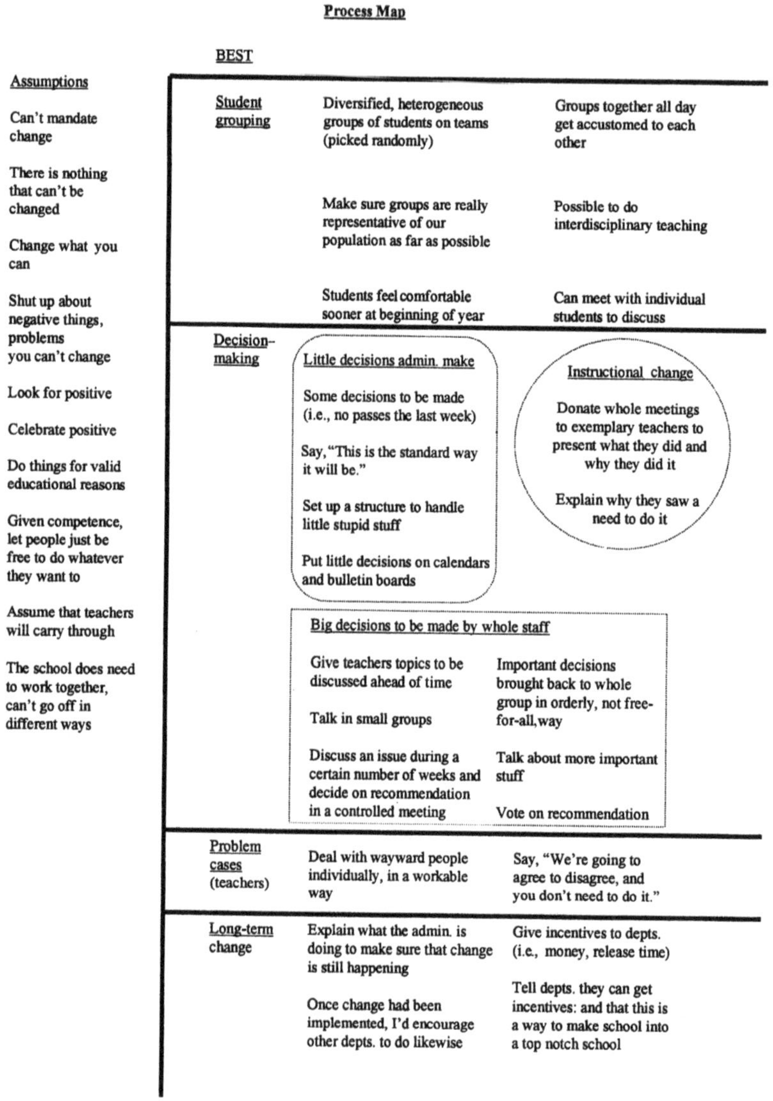

Process Map

BEST

Assumptions

Can't mandate
change

There is nothing
that can't be
changed

Change what you
can

Shut up about
negative things,
problems
you can't change

Look for positive

Celebrate positive

Do things for valid
educational reasons

Given competence,
let people just be
free to do whatever
they want to

Assume that teachers
will carry through

The school does need
to work together,
can't go off in
different ways

Student grouping	Diversified, heterogeneous groups of students on teams (picked randomly)	Groups together all day get accustomed to each other
	Make sure groups are really representative of our population as far as possible	Possible to do interdisciplinary teaching
	Students feel comfortable sooner at beginning of year	Can meet with individual students to discuss

Decision–
making

Little decisions admin. make

Some decisions to be made
(i.e., no passes the last week)

Say, "This is the standard way
it will be."

Set up a structure to handle
little stupid stuff

Put little decisions on calendars
and bulletin boards

Instructional change

Donate whole meetings
to exemplary teachers to
present what they did and
why they did it

Explain why they saw a
need to do it

Big decisions to be made by whole staff

Give teachers topics to be discussed ahead of time	Important decisions brought back to whole group in orderly, not free-for-all way
Talk in small groups	
Discuss an issue during a certain number of weeks and decide on recommendation in a controlled meeting	Talk about more important stuff
	Vote on recommendation

Problem cases (teachers)	Deal with wayward people individually, in a workable way	Say, "We're going to agree to disagree, and you don't need to do it."
Long-term change	Explain what the admin. is doing to make sure that change is still happening	Give incentives to depts. (i.e., money, release time)
	Once change had been implemented, I'd encourage other depts. to do likewise	Tell depts. they can get incentives: and that this is a way to make school into a top notch school

Figure 5.6.   Process concept map for third teacher.

good faculty, the use of student role models through integrated classes, and working on Paine's reputation in the community.

This teacher, a gifted classroom veteran who repeatedly noted her joy that she was not an administrator and therefore didn't have responsibility for the change process, struggled over the process map (Figure 5.4). She divided her ideas roughly into two parts:

1)   "I don't have an answer," given the constraints of teaching at Paine.
2)   "Stop making excuses and do something."

Included in her list of problems with the change process were several facts about life at Paine: it's simply easier for teachers to do what they've always done; everyone has too little time and feels overextended; removing good teachers from class to engage in change activities causes problems; new ideas generate more work for the same few people; and meetings this year have been frustrating and ineffective. In a poignant moment she commented, "Sometimes I long for the day when the principal . . . said, 'This is what you're going to do,' and you just did it."

The second part of her process map—the "stop making excuses and do something" part—listed four assumptions about how to proceed: accept that change is a slow process; get everyone—teachers, parents, students—to participate; take advantage of different people's strengths and energies; and provide time during the school day to work on school change.[1] Her process for creating an effective school had two activities: meetings of small groups to surface ideas; and a large group process for working with them. Her map also has a special place for those people who choose not to participate in change: they give up the right to complain.

The third pair of maps is the most complicated. The goals map (Figure 5.5) differs from the first two in its fairly detailed focus on students and curricular issues. In this teacher's opinion, the current situation allows or may even encourage students to make bad choices—minimal expectations, limitations due to early selections, course selection based on teachers rather than on content, and the presence of certain teachers with bad reputations. Because of the administrators' concern for safety, their current actions relate to behavior, rather than to curriculum and learning. The building's five broad Outcomes Accreditation (OA) goals[2] are included (critical thinking, improving English, improving math, improving self-esteem, and respect), along with the notion that such goals should be measurable. The upper right corner of the map presents ideal goals: preparation for the future; learning with an awareness of the process;

---

1   The teacher noted the tension between, on the good side, providing teachers time out of their workday for change efforts and, on the bad side, the problems of doing so.

2   The North Central Association's Outcomes Accreditation process is a five-year action research-like process. Schools are required to set goals (including at least one cognitive and one affective), collect baseline data related to achievement in these areas, put interventions in place for improvement, especially for those most in need, and finally to collect data to see if improvement has taken place.

purposeful student course selection; and student confidence that they *will* learn. Below this, is this teacher's commitment to equity: *All* students will be prepared well and similarly, using the same high standards. The curriculum development process under this couples administrative involvement with teacher collaboration. The final box points to the outcome of this collaboration. Students behave because they are actively involved in the classes they have purposely selected.

The related change process map (Figure 5.6) has five sections: a set of ten assumptions about change; an envisioned best process in four categories (student grouping, decision-making, teacher problem cases, and long-term change issues); the contrasting current situation in three categories; a list of current administrative actions; and a brief note on the district context. The ten assumptions point to a number of issues discussed in the literature on school change: the importance of collaboration ("the school does need to work together") and professionalism ("change what you can," "do things for valid educational reasons," "assume that teachers will carry through"); a focus on the positive; and a sense of free rein ("there is nothing that can't be changed").

In the next section, the best case process groups students heterogeneously; the current process causes problems due to fairly homogeneous student teams. Best case decision-making distinguishes among three categories of decisions:.

1)   "Little decisions" ("little stupid stuff") are the domain of the administration and should simply be made.
2)   Instructional change should be the domain of exemplary teachers who discuss their practice.
3)   "Big decisions" are processed first in small groups, then voted on after an orderly whole staff discussion.

Current decision-making is stymied because of pointless, ineffective meetings and meetings that include everyone but actively involve only a few individuals. In addition, in the current situation teachers may be told about issues when it is too late to do anything about them. The best case way to handle teachers who are not behaving professionally is to work with them individually, perhaps excusing them; in the current situation people are simply aware of who is not implementing certain policies, but there are no consequences (except perhaps being "bawled out" along with everyone else at a faculty meeting). The best way to effect long-term change is for the administration to make clear how it is supporting the change process and to provide incentives to departments.

The lengthy list of current administrative actions (Figure 5.7) documents an administration "putting out trees that are on fire, instead of looking to the future and planning." In this teacher's opinion, they deal with small matters, concentrate on the worst, sometimes fail to follow through, and may even revert to authoritarian methods in some situations. The district context is not helpful. A small circle in the bottom right-hand corner points to the control exerted from the central administration.

An initial processing discussion. Once the three teachers had finalized their maps, we faced the challenge of how to promote meaningful discussion of their

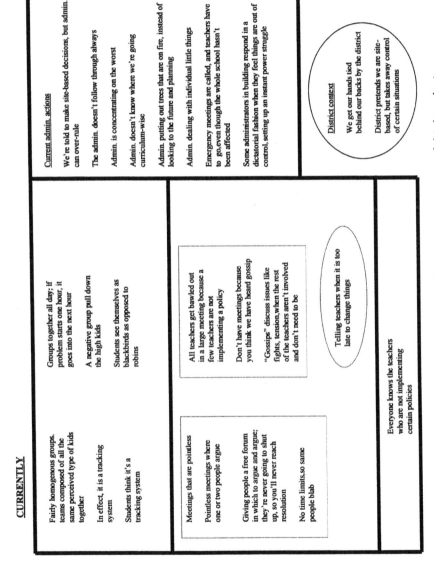

CURRENTLY

Fairly homogenous groups.
teams composed of all the
same perceived type of kids
together

In effect, it is a tracking
system

Students think it's a
tracking system

Groups together all day; if
problem starts one hour, it
goes into the next hour

A negative group pull down
the high kids

Students see themselves as
blackbirds as opposed to
robins

Meetings that are pointless

Pointless meetings where
one or two people argue

Giving people a free forum
in which to argue and argue;
they're never going to shut
up, so you'll never reach
resolution

No time limits, so same
people blab

All teachers get bawled out
in a large meeting because a
few teachers are not
implementing a policy

Don't have meetings because
you think we have heard gossip

"Gossips" discuss issues like
fights, tension, when the rest
of the teachers aren't involved
and don't need to be

Telling teachers when it is too
late to change things

Everyone knows the teachers
who are not implementing
certain policies

Current admin. actions

We're told to make site-based decisions, but admin.
can over-rule

The admin. doesn't follow through always

Admin. is concentrating on the worst

Admin. doesn't know where we're going
curriculum-wise

Admin. putting out trees that are on fire, instead of
looking to the future and planning

Admin. dealing with individual little things

Emergency meetings are called, and teachers have
to go, even though the whole school hasn't
been affected

Some administrators in building respond in a
dictatorial fashion when they feel things are out of
control, setting up an instant power struggle

District context

We get our hands tied
behind our backs by the district

District pretends we are site-
based, but takes away control
of certain situations

Figure 5.7. Current administrative actions (part 2 of process concept map of third teacher).

content—their similarities, which suggested points of consensus, and, more importantly, their differences—among a larger group of Paine faculty. To practice the process of integrating concepts across maps, we sponsored such a discussion at one of our Wednesday PDS brown-bag lunches, providing each staff member in attendance colored copies of the three maps, then asking them together to find similarities (if any) and differences.[3] One person characterized the meeting as a "science fair," as each person had six different maps in front of him or her (a goals and a process map for each of three teachers) and had to keep looking across them during the conversation. Figure 5.8 presents the graphic outcome of this discussion, which was a heated and often intense exchange that carried well beyond the time teachers typically stay. The group fairly quickly ended up discussing the change process, rather than goals; the current organizational tension at Paine is less over eventual outcomes and more over how to achieve them.

What did the group discuss? The brown-bag discussion group clearly agreed with those ideas present on at least two of the maps:

- The continuing problem of both staff and administration being overextended and forced to focus on immediate issues rather than broader, long-term issues
  ("We're always doing the immediate task"; "The problem is we are always playing catch-up and doing the immediate task"; "We do crisis management in this building, over and over and over . . .")
- Numerous problems with all-faculty meetings that year
  ("If you're going to start a meeting, have a time to start the meeting and start it"; "You hear seven people talk and that's always the same seven, and the rest just shut up, hoping we can get out soon"; "There is the thing, too, that—heavens forbid—you don't have something to talk about and you could cancel the meeting . . . instead of all coming down there and sitting while we hem and haw.")
- The recognized need for a process to discuss broader change issues
  ("We don't have a process to bring up new things . . . This is a credible question to ask: What's the process? Where does it go?"; "I feel so bad about new ideas in this building. I feel so bad because there's no processing time. There's no process to bring up an idea or a forum to bring up the idea"; "Why isn't there a forum in our building or a procedure or a process that would've allowed something like that to develop—a task force, a whatever [to discuss a proposed change]?")
- Memories of an effective past process
  ("Once upon a time we used to have . . . faculty council groups, and there were like six or seven people on each one and you actually met and just talked as a small group and you raised your concerns and those took the

---

3 The principal was scheduled to attend this meeting, but sadly, from our perspective, an unscheduled visit from the superintendent of schools prevented that. There were no administrators in attendance at the discussion.

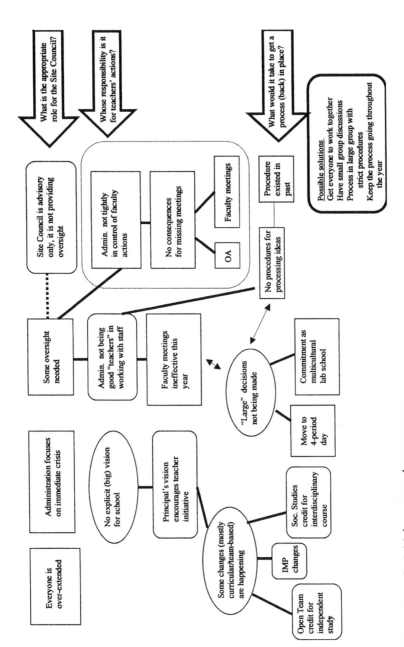

Figure 5.8.   An initial consensual concept map.

place of one of the meetings because you actually got to talk about things and then propose things . . . and make different solutions, and then the representatives from those got together and they prioritized things and that came to the whole faculty meeting.")

- Suggestions for a better process for discussing change activities ("Can't we run the faculty meetings like we know [how] to run our classes? Couldn't we have small groups of us meeting together and listen to each other and then get back together again?"; "Just common sense, good cooperative learning for adults . . . when you really want 'em to talk and give 'em a task to do, pros and cons, and come out with that and then come back to the whole group and do a quick report.")

Several points important to the change process in the school emerged. First, the group agreed that at that point there was no explicit vision for Paine that the faculty commonly understood, a fairly damning comment, given the literature on school change. Those who had taken part in the original restructuring at Paine recalled a mission statement developed in 1988. People noted that the current principal's vision clearly encouraged teacher initiative and that, as a result, some changes—especially team-initiated curricular changes—were occurring. Teachers gave three examples of such change: 1) one team's getting course credit for independent study; 2) the math department's work with an innovative mathematics program; and 3) one teacher's getting students social studies credit in an innovative interdisciplinary humanities course.

However, there was equally clear consensus that, by contrast, "large," school-level decisions were simply not being made, a second important point. Administrators were blamed for not acting like teachers; i.e., for not using the techniques of effective "teaching" with their faculty. As one teacher put it:

Part of the problem maybe is that the administration has not been teaching for a long time, and they don't know how to teach anymore or hold a group together anymore . . . They're reverting back to what they've seen people do before, and that's always never worked, but it's what they've seen before, and, so okay, we'll get an overhead and we'll read it to you.

Teachers gave two examples of what they perceived to be poorly processed non-decisions: 1) whether or not to move to a four-period schedule; and 2) whether or not to become a multicultural lab school for the district. Because of ineffectual faculty meetings and the lack of a process for discussing new ideas, faculty expressed concern about the outcome of these important decisions, which were still in process.[4]

A third point related to possible solutions for this problem. Several in the group recalled the procedure used in 1988–1990 when Paine first ventured into

---

4 Two years later, neither the four period day nor the laboratory school commitment has been implemented.

restructuring, and others added ideas about how such a renewed process might work. However, the question of *how* to get something like it reinstated remained unanswered: "Who says we're gonna change our process? Who says we're gonna regroup into it?"

Finally, the group briefly touched on a central question for a restructuring school: Exactly who should be responsible for teachers' actions—the administration or the teachers themselves? Teachers agreed that someone had to provide oversight for the change process and that, at this point, the Site Council with its advisory role was not doing that. Some teachers blamed the administration for not being in tight control of the faculty, noting that there are no consequences for missing faculty or committee meetings. But another teacher said," So whose responsibility is it for us to control our behavior? Is it our responsibility? Is it somebody else's?" In the overriding discussion of how to process new information, this point was not emphasized, although the same individual added—much later—"I call it accountability . . . professional accountability."

By the end of the discussion, we felt confident that the project's Phase One question had been answered affirmatively. As Khattri and Miles (1994) would have predicted—and as the transcribed tape documents—the three concept maps generated rich discussion about core issues related to change at Thomas Paine. This was good, however, but not good enough. Even before Phase One was complete, we realized that the mapping process we had used, i.e., university-based researchers conducting interviews that were recorded and fully transcribed—was costly and unrealistic for most schools, and especially for schools in urban centers. Phase Two emerged from this realization as we sought to develop a more cost-effective mapping process that extended what we had learned in Phase One.

The mapping intervention at Paine: Phase Two

As noted, the second phase overlapped with the first. It asked a slightly different question: To what extent can concept mapping be integrated into the ongoing process of change in an urban school? Phase Two rested on two assumptions. The first assumption was that more was better; i.e., if three maps generated good discussion, imagine the conversation an additional dozen or so might inspire. In order to have a more inclusive set of maps for the building, we expanded the mapping process to include two other types of individuals: staff (six teachers and two non-certified staff) who were suggested by a planning committee and selected by the principal in October; and the four members of the administrative team. Ranging in age and teaching experience within and outside Paine, the teachers represented all of the school's academic teams. The two non-certified staff were extremely valued and knowledgeable resources, highly respected members of the school community who had not previously participated in PDS research activities. Aware that racial issues affected the overall function of the school, this round of interviews targeted three African-American staff, and an African-American researcher conducted these interviews.

Furthermore, we knew that one of the problems with collecting information about change processes from building staff was that administrators in general—and the building principal in specific—end up as unavoidable targets, often perceived as the "bad guys" limiting change. This phase, then, had an important additional purpose: to allow administrative staff to chart *their* vision of Paine's outcomes and a process for reaching them, in contrast to the non-administrative staff. In order to promote meaningful dialogue, administrators' perceptions of the ways in which teaching faculty were affecting change in the building needed to be equally and clearly stated.

Phase Two had a second framing assumption. If the mapping process were to be truly viable in settings where university collaborators were unavailable, it should be home-grown rather than completely reliant on experienced researchers. Our solution—which, admittedly, only partly addressed this concern—was to use seven graduate assistants who played various roles at Paine (five of whom were former teachers) to conduct interviews, create maps, and participate in a year-long discussion of the mapping process and its results.[5] The graduate assistants had all taken at least one course that covered interviewing, several had previous research or evaluation experience, and during the second year of the project we met as a group almost weekly to increase the comparability of the maps.

By the middle of that second year, all fifteen individuals we had targeted (four administrators, three teachers from Phase One, six teachers from Phase Two, and two non-certified staff) had been interviewed, a goal and process map prepared for all but one—a total of 29 different maps—and the individuals had, with one exception, confirmed their map's accuracy.[6] Our decision to expand the number of individuals developing maps unavoidably led to problems in comparison. In our effort to simplify the method, not all of the interviews were taped or transcribed. Some Phase Two maps were extremely simple compared to the three drawn during Phase One, and it was hard to know to what extent the individual and/or the interviewer had altered the ideas when putting them in map form. Because this was a qualitative process, the "member checks" were extremely important. If the person interviewed believed that a fairly simple map captured his or her thoughts, then that map was added to the collection; the map was edited as necessary until the individual accepted it as an accurate portrayal of what he or she had said in the interview. The resulting maps varied

---

5 Ideally, Paine teachers would have interviewed and developed maps for their peers, but in the climate of the building that was not a realistic option; teachers simply couldn't add this work to their loads.

6 One teacher, who had been interviewed in the spring of year one, moved to another school in year two. He was not interested in continuing involvement at Paine, so we did not use his maps during the second year. A second individual, one of the non-certified staff, did not feel he had the knowledge to create a goals map; he viewed that as the professional staff's job, while his was to provide support to the building.

greatly in level of detail, and, given our framework of organizational learning, we faced the challenge of how to add them to the learning process of the building. (See Figures 5.9 through 5.12 for a sample of the range of maps.)

We eventually settled on two strategies. First, we developed a process for a whole faculty refinement of the goal maps' content and a discussion of change processes. The university people who had created the maps identified individual concepts that appeared in the collection (e.g. "relevant and current curriculum," "trust," "improved attendance"). With the maps spread on a table in front of us, we refined this list of concepts by checking, one at a time, that all of each person's goals were on the list. If a concept was not yet there, it was added to the list; once it was there, it was not repeated when it appeared on another map. We then created a strip of paper for each goal, a total of 30 or so that were put in envelopes. Interestingly, concepts related to diversity appeared on almost everyone's map, and the only individual who explicitly included the term *racism* was not African-American.

Concept Map

<u>GOALS</u>

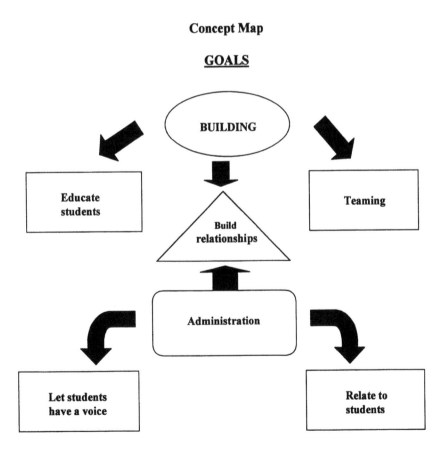

Figure 5.9. Phase two sample goals and concept map.

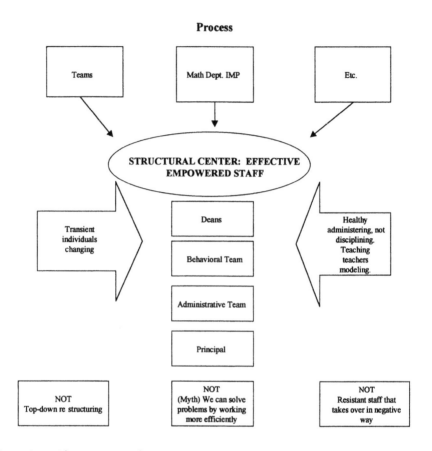

Figure 5.10. Phase two sample process map.

At an all-school PDS forum in January of the second year, seven active PDS faculty volunteers facilitated small group discussions with roughly ten colleagues each. Their task was as follows: place the goal strips on a table, and select (after conversation) what the group believed to be the five most important goals for Paine; rank order these five from the most to the least important, and arrange them on a poster board; then, in the time remaining, discuss the "best way to implement meaningful change" at Paine. A graduate assistant served as recorder for each group, and the discussions around the various tables appeared lively. By the end of the afternoon, the groups had created seven maps of what the Paine staff believed to be the most important goals for the school and a list of suggestions for implementation (e.g. "more academic electives," "alternative programs for the disruptive students in the school"). Some groups "cheated" (as they put it) by grouping concepts under headings; others produced a straight-forward, ranked list. Many examined other groups' poster/maps with extreme interest.

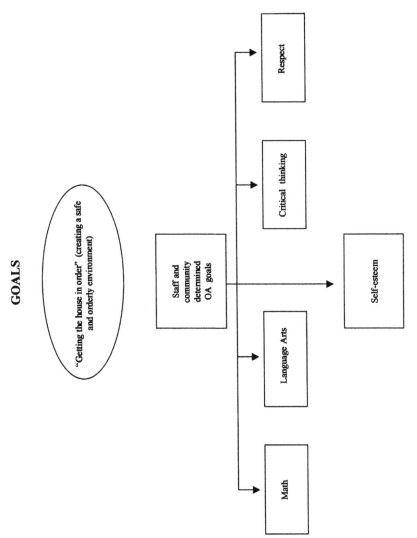

Figure 5.11. Phase two sample goals and concept map.

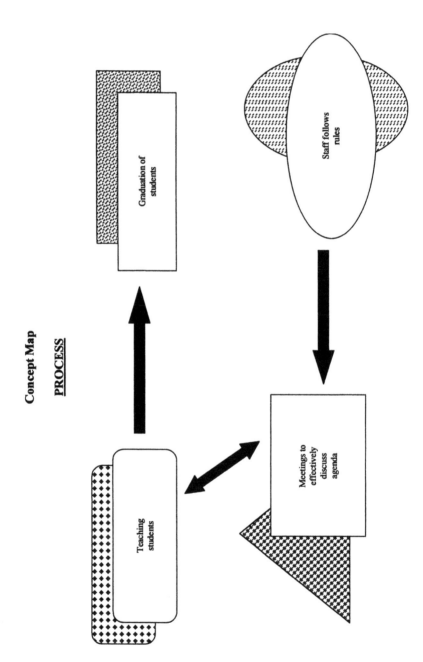

Figure 5.12. Phase two sample process concept map.

The following Wednesday, the PDS brown-bag lunch featured a discussion of the seven groups' work. The poster boards were placed around the room, and the 20 or so people in attendance analyzed similarities and differences in an effort to arrive at a small number of goals that could be labeled top priority for the school. In a surprisingly short time, the group came to a consensus on three goals that were common to all seven poster boards:

- higher expectations for both students and teachers
- motivating or empowering students to learn
- a more relevant curriculum

How to implement these goals became the topic of February's all-school PDS forum, where, in eight small groups, faculty and staff discussed two questions: "Assuming we really wanted to achieve these outcomes at Paine, what process would be appropriate, and what barriers would have to be overcome?" Again, people appeared intent during their conversations, and the groups generated 53 ideas for the change process. These are categorized in Table 5.1.

The list of potential barriers was far shorter—24 items—but clearly a major impediment to achieving the goals. Roughly a third of the barriers related to resource issues (e.g. money, lack of staff energy, time); another third related to the existing status at Paine (e.g. lack of a common prep time, mixed messages to students, opposition within the building); and the remaining barriers related either to Paine's students (e.g. transient student population) or the school's context (e.g. poverty, racism). These typed lists of potential changes and barriers, developed collaboratively by the Paine faculty at two meetings, were then given to the administrative team, to a teacher representative on the Site Council, and

Table 5.1. Ideas for the change process.

Category	Number (%)	Examples
Specific structures or activities for students	30 (57%)	• Tutorial hall/hour for all students • Better regular ed/special ed mix • Every kid has an adult to check in with • Campaign for better hall behavior
Changes in teaching/teachers	9 (17%)	• Team strategies • Following existing rules; staff-wide consistency
The change process itself	8 (15%)	• Consistent policy- and decision-making • Unstructured in-service days
Parental and community involvement	3 (6%)	• Increased involvement of parents, businesses, teams
Outcome statements	3 (6%)	• Motivation

to the PDS Steering Committee for their consideration. This, then, was our first strategy for integrating the concept maps' content into ongoing discussions at Paine. It sought to process the variety of concepts people had mentioned, allowing the entire faculty and staff to choose among these ideas and develop priorities as a group.

The second strategy was purposefully less inclusive, primarily targeting the administrative team. It sought commonalities in order to contrast administrators' and teachers' views of goals and change in the building. Taking advantage of the university person-power available to Paine, we developed drafts of two summary maps that combined goals and process: an administrative team map based on the four administrators' maps (Figure 5.13) and a Paine teachers' map (Figure 5.14) based on the eight teachers' maps. In April of the second year, the teachers who had been interviewed reacted to the drafts at a brown-bag discussion, and the administrators did likewise at a meeting devoted to this topic. It is important to emphasize that these maps were created as discussion triggers, not as research documentation; they did not purport to fully represent the content of each and every source map, but rather they combined common elements into a unified whole. Because the teachers as a group and the administrative team accepted the maps as adequate summaries of the individual maps' content, we deemed them "good enough" for the purpose of comparative discussion.

What did people note in their comparisons? Both summary maps included common goals for the entire school and student learning as a central focus (Figure 5.15). There was essential agreement on student goals (the details of which were omitted from the teachers' map due to space constraints). There was a clear contrast, however, between the administrators' roles in these maps. In the teachers' map, there is a division between administrative and participatory decision-making. The administration makes "little decisions" and decisions in times of crisis; the staff makes decisions related to curriculum and instruction and to "policy—big issues" through a participatory process that includes committees, departments, individual teacher and student leaders, and the building's Site Council. On the participatory side, "monitoring" and "whole school accountability" are included en route to improved student learning. In the administrators' map, the administrative team is responsible for three things: staff support and supervision; creating a safe and stable learning environment; and getting resources. Through a variety of activities, including developing "care, respect, and trust" and "leadership in every staff member," the team's functioning leads to an "effective, empowered staff." The decision-making process is not made explicit, except to note that it must be open and "hard issues" discussed. Although the administrators surely believe in accountability, it does not appear as a concept on the map.

In April of the second year, the administrative team (plus a then-dean who has since become an assistant principal) spent time examining the summary maps and reviewing the mapping process. To facilitate comparison of the individual administrators' maps, we made a list of the concepts present in each of them and recorded how many of the other maps included that concept. Only

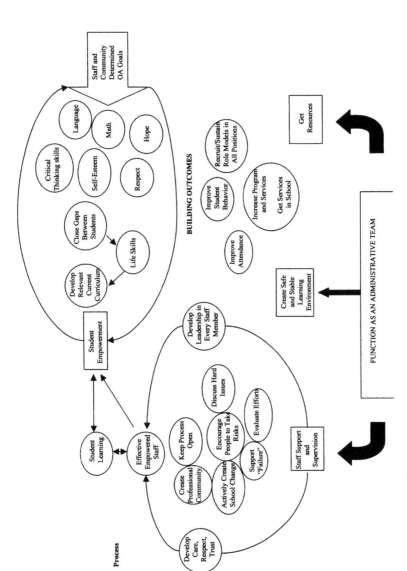

Figure 5.13. Administrative team concept map.

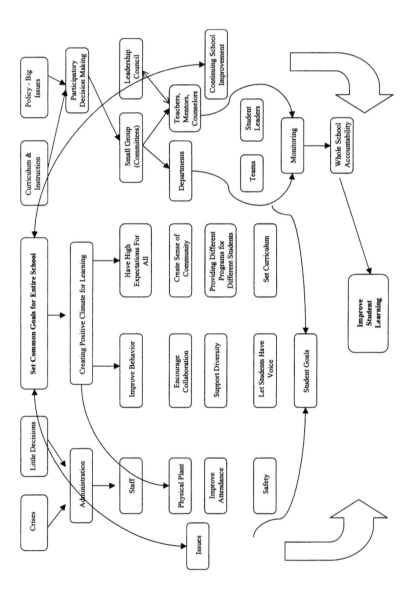

Figure 5.14. Teachers' concept map.

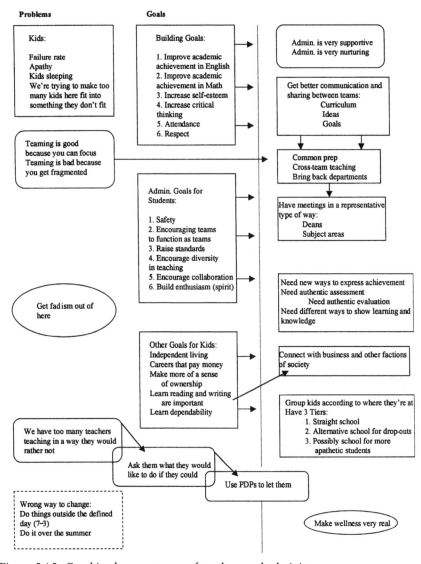

Figure 5.15. Combined concept map of teachers and administrators.

two concepts ("raising attendance" and the "bank partnership") appeared on all four maps; those appearing on three maps included "student achievement," "professionalism," "an empowered staff," and "teachers/staff" in general. Focusing on the summary maps, the administrative team discussed three topics:

- An initial comment was to note that empowerment was a central feature of the administrative team map, but not included on the teachers'. One AP noted, "Empowerment is so new, it scares people . . . empowerment's more dramatic than just the word." The principal agreed: "This school wanted a knight to come in and save it, [but] at the ground level, a strong, empowered community is what will make the difference."
- Another administrator pointed out the amount of space devoted to student outcomes on the administrative map vs. the amount of space devoted to decision-making processes on the teachers' map. "If we are student-oriented, we may not nurture the staff enough." He continued that the administrators need to do that so that people "can bring hard issues and challenges up; that's when I've seen results."
- The final topic was the value of the concept maps as a "learning process for principals," as something "you could put into a portfolio." Administrators mentioned specific programs they might want to track, and one explicitly commented on their changing role in a building that was no longer perceived to be out of control: "We've been kicking ass so long, now we're not kicking ass. So what do we do? How do we define our new role? We're on a new page here . . . We could use concept maps to see if we've followed through on what we wanted to do."

Phase Two of our work asked the question: To what extent can concept mapping be integrated into the ongoing process of change in an urban school? The two strategies—a general faculty discussion of the content of the goals maps and a needed change process; and a targeted discussion of consensus maps—surely integrated the mapping process into Paine's ongoing change effort, although it was not clear to what extent the ideas would be reflected and acted upon over time. In retrospect, our Phase Two assumptions—that many maps would be more useful than a few and that the process could be made more home-grown—seemed incorrect. Fifteen qualitatively developed maps created far more information than a group of practicing teachers could readily process, and, even during Phase Two, the mapping effort relied almost exclusively on the efforts of part-time university staff or students working in the school. In the spirit of action research, however, we surely came to understand the mapping process and what we might do differently in a further project, which is the subject of the next section.

Lessons learned from the mapping process

During its two phases, this project taught us on one level that concept mapping could be a viable tool for fostering discussion of change issues in an urban school and that the mapping process could be integrated into a school's ongoing change process. Concept mapping, then, can be one useful tool in an array of techniques for organizational learning. The process encourages people to make visible the concepts that guide their practice, and it facilitates comparison and contrast across individuals and roles. Our experience suggests that a qualitative approach to map development, while it may not generate completely valid representations of people's views of reality, is nonetheless sufficiently robust to generate thoughtful discussion—which is what we intended. However, while the effort at Paine surely demonstrated this point, our reflections following the two years point to a metaphoric glass that looks half empty despite some evidence that it is half full. The real question remaining is whether it was the concept maps *per se* or rather the opportunity for extended interviews and discussion of the change process that made a difference. These thoughts lead us to three final points.

First, the act of qualitatively developing concept maps from interviews is nontrivial and not to be taken on casually. In Phase One when the interviews were recorded, the process of identifying concepts or phrases in the transcriptions took a minimum of one and a half hours, apart from transcribing the tapes—of itself a time-consuming task. Even then, it was not always clear that all of the ideas had been identified. The Phase One process of putting these onto post-it notes took time, as did the second and third meetings with the teachers. Adding in the time it took to type the concept map on the computer, each pair of the three Phase One maps required between six or seven hours of researcher/facilitator time to generate, plus roughly an hour and a half of a busy teacher's time. Additional time was needed for analysis and discussion of the maps.

We sought to remedy this in Phase Two by using a number of map drawers, by not taping all the interviews, and by eliminating the post-it note stage and directly creating draft maps on the computer. As was noted, however, this created other problems, most importantly the comparability of maps. In our effort to make the process easier by adding multiple map drawers and removing transcription, we reduced the consistency of the process. We rationalized that this was all right—i.e., that the single criterion that mattered was that each map must satisfy the individual whose ideas it contained—but were left with a nagging fear that ideas may have been lost in the translation. As our process evolved, we became increasingly aware that our mapping was less a research technique generating accurate data about Paine, as some approaches to organizational learning would require, and more a way to capture people's perceptions so these could be discussed publicly. We concluded that the mapping process is never easy, even when made (more) simple, and it is not finally clear to us whether or not the act of creating concept maps was worth the substantial investment of our limited resources. In retrospect, Trochim's highly quantitative and computer-based approach that systematically generates group maps took on increased appeal.

Second, we now believe our plan included far more maps than were necessary to foster meaningful conversation about change, especially among a faculty actively reconstructing their school over a period of years. Most people reported that they were honored to be asked to participate and that they enjoyed the process; several were amazed at the time we spent getting *their* ideas into the form *they* wanted—a perceived benefit for the individuals involved. This is surely evidence of what Patton (1997) would label the process use of evaluation, but it is not clear if this benefit outweighed the cost of developing the maps. The intensity and richness of the lunchtime conversation during Phase One made us wonder, after the fact, whether we needed to create twelve others. Had we taken those three initial maps and added the four administrators' maps, we could still have held a comparative discussion that would have surfaced important issues. What matters, we now believe, is the reflection on information that is good enough to trigger a thoughtful discussion, and a structure and the ability, then, to act on it.

Finally, the parallels between our mapping experience and the evaluation use literature may help explain our lingering concern that the two strategies we used in Phase Two may finally have had little impact. Researchers studying evaluation use noted many years ago a distinction among three types of use:

- instrumental, where evaluation information leads directly to specific actions
- persuasive, where an individual uses evaluation information for some personal end
- conceptual, where evaluation information is added to the store of available materials people keep in their heads from that point forward (Leviton & Hughes, 1981)

In recent research, McCormick (1997) has added a fourth type of use, which she calls "processing use," during which an individual packages evaluation information for distribution or discussion purposes; some might label this a form of dissemination. The efforts we made in Phase Two were exactly of this nature. In one strategy, we created strips of concepts and structured times for faculty discussion of them and of products developed from them; in the other, we created summary maps and structured time for the administrative team to discuss them. Beyond this processing, we don't have evidence of the extent to which Paine faculty or administrators used the results of the discussions to make specific changes or to persuade others, or even if the maps' content or the processes for discussing it somehow became part of or altered people's mindsets. This clearly remains a question for additional study.

References

Barenholz, H. & Tamir, P. (1992). A comprehensive use of concept mapping in design instruction and assessment. *Research in Science and Technological Education* 10 (1): 37–52.

Hart, A. & Bredeson, P. (1995). *Toward a theory of professional visualization.* Paper presented at the annual meeting of the American Educational Research Association, Chicago.

Khattri, N. & Miles, M.B. (1994). *Cognitive mapping: A review and working guide.* Sparkill, NY: Center for Policy Research.

Leviton, L. & Hughes, E.F.X. (1981). Research on utilization of evaluations: A review andsynthesis. *Evaluation Review 5* (4): 525–548.

McCormick, E. (1997). *A conceptual framework for the study of evaluation use.* Unpublished manuscript. University of Minnesota.

Margulies, N. (1995). Map it: Tools for charting the vast territories of your mind. *Interactive Comics, Volume I.* Tucson: Zephyr Press.

Morine-Dershimer, G. (1991). *Tracing conceptual change in preservice teachers.* Paper read at the Annual Meeting of the American Educational Research Association, Chicago.

Novak, J. (1990). Concept mapping: A useful tool for science education. *Journal of Research in Science Teaching 27* (10): 937–949.

Oldfeather, P. (1994): Drawing the circle: Collaborative mind mapping as a process for developing a constructivist teacher preparation program. *Teacher Education Quarterly 21* (3): 15–26.

Patton, M.Q. (1989). A context and boundaries for a theory-driven approach to validity. *Evaluation and Program Planning 12:* 375–377.

Patton, M.Q. (1997). *Utilization-focused evaluation: The new century text.* Thousand Oaks, CA: Sage Publications.

Posner, G. & Rudnitsky, A. (1986). *Course design: A guide to curriculum development for teachers.* New York: Longman Inc.

Trochim, W. (1989). An introduction to concept mapping for planning and evaluation. *Evaluation and Program Planning 12:* 1–16.

Wandersee, J. (1990). Concept mapping and the cartography of cognition. *Journal of Research in Science Teaching 27* (10): 923–936.

6

Organizational Consequences of Participatory Evaluation: School District Case Study[1]

J. Bradley Cousins

Introduction

In the present study, I consider the potential for collaborative or participatory evaluation as an organizational learning system by examining its consequences at the school and district levels. A case study that employed multiple sources of data collected at different points in time is used to accomplish this goal. This study not only provides the opportunity to understand the organizational consequences, planned and unplanned, of collaborative evaluation, but also provides a chance to identify and track the influence of conditions and factors that led to them.

The organizational learning literature is burgeoning. While the bulk of work in this domain has been carried out in domains of study external to education,

1 This research was funded by a grant from the Social Sciences and Humanities Research Council of Canada (No. 410–92–0983). A previous version of this paper was presented at the annual meeting of the American Educational Research Association, San Francisco, April 1995. The views expressed in this paper belong to the author and do not necessarily reflect those of the Council. The author is indebted to Liane Patsula, Margaret Oldfield, Cheryl Walker and Carolyn Brioux for their assistance with aspects of the data collection, processing and analysis. The person referred to in the paper as the research committee chair was instrumental in coordinating data collection.

recent connections to educational research and practice have been made (Cousins, 1996a; Louis, 1994; Louis & Simsek, 1991). Fundamental to conceptions of organizational learning is the development among organization members of shared mental representations or understandings of the organization and how it operates. Some theorists take the view that such learning need not be directly observable through changes in organization members' behaviors. What is important is that the range of potential behaviours is changed (Huber, 1991). Others, however, insist that it is through behaving and experiencing that we learn and that organizational learning is not possible unless a process that allows for error detection and correction is operative (Argyris, 1993; Argyris & Schön, 1978). Most theorists agree, however, that organizations experience different levels of learning ranging from low-level, first-order, or single-loop learning, where change is incremental, to high-level, second-order, or double-loop learning, where fundamental assumptions about the organization and its operation are surfaced, questioned and ultimately altered (Fiol & Lyles, 1985; Huber, 1991; Lant & Mezias, 1992; Lundberg, 1989).

Daft and Huber (1987) differentiate between system structural and interpretive organizational learning systems and propose that organizations oscillate between the two, depending on circumstances, requirements and needs. System structural learning systems assume that a given organization exists in an objective environment, that understanding leads to action and that the rational analysis of data is how that understanding is to be achieved. On the other hand, interpretive systems assume that the organizational system gives meaning to data which are inherently equivocal and that learning is a consequence of dialogue and the development of shared interpretations among organization members. The two systems are not thought to be incompatible or mutually exclusive.

Key to most conceptions of organizational learning is the concept of social interaction. Shared interpretations of ambiguous phenomena are arrived at through dialogue and deliberation among organization members. Louis and colleagues refer to this as "social processing" and identify interpersonal networks as being powerful organizational learning systems (Louis, 1994; Louis & Dentler, 1988; Louis & Simsek, 1991). It is this concept that provides the basis for a connection between program evaluation and organizational learning.

I define program evaluation as systematic inquiry carried out in support of organizational problem-solving and decision-making. While evaluation activities naturally involve judgements about the merit and worth of programs they also serve to help progam implementors to understand their programs and to determine reasonable courses of action for program improvement. While many program evaluations are carried out by trained (often external) researchers and consultants, there is growing interest in forms of evaluation that involve quite directly individuals responsible for program implementation and other stakeholders such as program developers, sponsors and even intended beneficiaries (Ayers, 1987; Cousins & Earl, 1995; Fetterman, 1994; Patton, 1994). My basic proposition is that evaluation projects that involve as collaborators or co-evaluators individuals responsible for program implemention and management have the potential to develop organizational learning capacity.

Evaluation activities can involve organization members in several ways. Members can simply be involved as informants or as providers of evaluation data. Alternatively, they can be more directly involved in actually carrying out the evaluation. In traditional stakeholder-based evaluation models, for example, program stakeholders are involved in scoping out the evaluation or otherwise shaping its design and objectives, and, subsequently, as aids to interpreting the findings. Such activities are likely to foster the use of evaluation findings, since they will ultimately make the evaluation more responsive to the needs of the organization stakeholders (Ayers, 1987; Bryk, 1983). In other forms of collaborative evaluation program implementors jointly participate with researchers in developing instruments, collecting data, analyzing and interpreting data and summarizing findings (e.g. Ayers, 1987; Cousins & Earl, 1995). In an even more extreme case, program implementors run their own evaluations and merely consult research experts periodically for advice (Fetterman, 1994). Each of these approaches provide opportunities for interpretive dialogue and deliberations (social interaction) that could support the development among organization members of shared understandings of the program and indeed the organization under study. For example, direct participation in evaluation might help organization members to develop reasoned understandings of organizational successes and failures or to understand the implications of changes in the organization's environment. Alternatively, organizational members may develop a richer appreciation of cause-and-effect relationships among components of the programs or services being implemented or their unintended effects (good and bad).

While the arguments for engaging program implementors in evaluation activities may have appeal, there has been little in the way of direct study of collaborative evaluation as an organizational learning system. Some recent studies have helped to develop our understanding of the conditions that support such activities in schools and school districts (Cousins & Earl, 1995), but by and large the consequences of these strategies have been restricted to the program under study. In the present study I examine the consequences of a collaborative evaluation that extended beyond the focus for evaluation. In particular, consequences for learning in the wider organizational context are examined.

The context for study
A form of participatory evaluation was implemented in a high school located in a medium-sized public school district in east-central Ontario. This school (one of eight high schools in the district) had received funding from the provincial Ministry of Education and Training for a pilot project to explore alternative modes of curriculum delivery for students in the "transition years," grades 7 through 9 (ages 11–14). The two-year pilot project was implemented in advance of Ministry policy changes that included, among many other initiatives, "destreaming" grade 9. Staff at the high school designed and implemented an innovative grade 9 program that ensured mixed-ability grouping and adhered to

the principle of minimizing for students the number of different teacher contacts while at the same time maximizing the time spent per day with students in the home group.

Pilot funding (about $120,000 Cdn) supported a full year of school-based planning and the first year of implementation. A steering committee consisting of representatives of the school's teaching staff, school and district administration, parents and the student body was created, and subcommittees, a research subcommittee among them, were formed to aid with the implementation effort. The research subcommittee consisted of about six teachers in the school, although its composition changed slightly as the project evolved.

As part of the requirements for funding, grant recipients were encouraged to draw on the services of an external monitor to help evaluate their programs. I was recruited by a senior administrator to fill this role. Over a two-and-a-half year period I worked part-time in partnership with the research subcommittee to design and carry out a comprehensive evaluation project.[2]

Collaborative decisions were made about project design, scope, data collection, analysis, and reporting. I was able to contribute significantly to the technical research needs of the committee but did so only on the understanding that committee members would be active participants in "doing" the research project. As the evaluation project unfolded, the committee members found themselves designing and adapting instruments, collecting survey data, content analyzing comment data, interpreting statistical output, and writing parts of the final report. The design was a pre-test/post-test non-equivalent control group arrangement with data collected from teachers, students, and parents.

During the first year of pilot implementation, several problems emerged. Not all of the teaching staff were philosophically sympathetic to the pilot. Some were reluctant to let go of the streamed system, since they felt that the adoption of mixed-ability classes would be a disservice to motivated and academically serious students, while others felt unprepared to teach mixed-ability classes. Finally, many viewed the pilot as being a divisive force in the faculty. The disunity introduced into the school by the pilot project raised serious questions about the costs and benefits of disturbing the status quo.

By the second semester a number of events served to add fuel to the embers of discontent. First, a string of administrative "glitches," such as scheduling conflicts, timetabling complications, and the like, created anxiety and frustration among staff and students. Next, a visiting representative from the Ministry, in response to hearing the views of dissenting staff, intimated that the project was a bit too unwieldy. Third, the grade 9 students—all of whom were necessarily participating in the pilot—were becoming more vocal about their own dissatisfaction. They perceived that the project was isolating them from the rest of the school. Finally, preliminary data from the evaluation study implied that the pilot was exacerbating the very problems it was intended to ameliorate.

---

2 I worked in this capacity on a non fee-for-service basis which was consistent with the mandate of the institute with which I was employed.

In response to these events and observations, the school administration—which had consistently enjoyed the support of the central board office senior administration and the latitude to make its own decisions relatively autonomously—came to a pivotal decision, one that was later to be regretted. The principal, with support from his administrative council, decided to hold a faculty-wide referendum concerning the disposition of the pilot project. By this time it was common knowledge that the Ministry of Education and Training planned to implement their transition years initiative (including destreaming) in one-and-a-half year's time. But the question posed to school teaching staff—for which the principal required a two-thirds majority affirmative vote—was: "Should the pilot project continue into the next academic year?" Given the sense of urgency with which the decision needed to be made, the vote was imprudently held days before the spring break. The final tally revealed a fifty-fifty split, an outcome that meant discontinuation for the pilot in the year to follow.

The previously supportive senior administrator was bewildered and encountered considerable pressure to provide an explanation as to the disposition of this politically visible pilot. As a result, the collaborative evaluation team, through me, received an invitation from the administrator to expand our mandate. The superintendent requested that we systematically investigate the reasons underlying the faculty's decision for discontinuation. We accepted and decided to piggy-back an end-of-year survey onto our post-test teacher data collection activities. The instrument we designed asked respondents to anonymously indicate how they voted and why.

Meanwhile, continuing analyses of the pre-post data collected from students, teachers, and parents were, by and large, turning out not to be supportive of the program. The research committee, through considerable dialogue and deliberation, came to understand these findings as a manifestation of the somewhat predictable planned change phenomenon known as the "implementation dip" (Fullan, 1991; Huberman & Miles, 1984). The research pressed on and the team, several members of which were program advocates, struggled with the prospect of having to report negative findings in a situation that was already politically volatile.

My role throughout the collaborative evaluation process was quite active. I provided advice on technical matters and helped design the study. I coordinated data processing and statistical data analyses, trained committee members in the skills required for content analysis and statistical output interpretation, and contributed heavily to the structure, format, and substance of the final report for which I was acknowledged as co-author. But the report production phase ran into several unanticipated (by me at least) delays. My efforts to motivate the collaborative team through my contact with the committee chair were only moderately successful and we were running short on time. Unfortunately, delays eventually meant that my involvement in the production of the final report had to be diminished somewhat due to competing demands on my time. Meanwhile, the previously supportive senior administrator conspicuously distanced himself from the project and its repercussions. At about the same time that we were compiling the report, he moved into the directorship of another Ontario school district located some distance away. His replacement, a superintendent promoted from within the organization, knew about the evolution of the pilot project and its evaluation and was supportive of it.

The evaluation committee chair persevered. Through successful negotiations involving his principal and the succeeding superintendent he managed to secure resources to free the evaluation team members from their teaching duties to work on the report and to keep the project moving ahead. The team was able to keep on task and went to great pains to ensure that the report was true to the data. All evidence, even with its oppositional tone, was included in the report. Precautions were taken to highlight the study as a formative exercise, one that would be beneficial to planning future transition years' curriculum re-design efforts. With considerable effort and consternation, the much-awaited final report was finally released about four months overdue and seven months after the pilot project had been terminated.

The final products were an 80-page final report with technical information appended and a five-page executive summary. Three hundred copies of the full report were printed and distributed to all staff at the high school, administrative staff in other schools, the board of trustees and senior administration, and more broadly in the province to educators in other school districts. The five-page executive summary was circulated much more widely within the local school district. Finally, several other presentations were made by the committee chair and the principal to the local board of trustees, regional transition years' conferences, and various school boards around the province. Some of these presentations took place prior to the release of the final report.

After the document had been released, it was necessary for me to turn my attention to other priorities. The evaluation team—though members participated in various dissemination activities—essentially returned to their normal teaching duties breathing a collective sigh of relief that the two-and-a-half year research project was finally complete.

Case study methods
I drew on four sources of data for the case study. First, throughout the duration of the research project I kept participant observation notes in the form of records of telephone calls, meetings, and other exchanges and interactions. Some of these were handwritten. Others were audiotape-recorded dictations by me that were subsequently transcribed verbatim. In all, I generated about 40 pages of field notes. Secondly, I re-examined teacher, student, and parent data from the evaluation project itself. These were in the form of both quantitative questionnaire responses as well as written comment data taken from returned surveys.

A third source of data was a post-intervention survey of teachers in the district, those belonging to schools offering intermediate (grade 7–10) or senior (11–graduation) programs and, therefore, having a stake in the transition years policy initiative. The survey took place over one year after the evaluation was released. It was part of a larger study of teachers' attitudes toward local applied research and asked respondents to provide views about the transition year's study as well as general opinions about evaluation and applied educational research. We received quantitative and written comment data from 178 teachers and principals.[3]

A final source of data was interviews with school and district staff with a reasonable knowledge of the evaluation project and/or its consequences. The interviews took place two months after the survey. The nine participants were members of the post-intervention research committee (chair and two teachers), the steering committee, teaching and administrative staff at the high school, and supervisory staff at the central board office. The interviews were conducted by myself and two trained assistants in a concerted two-day effort. We each used a standard interview guide and subsequently summarized the audiotaped interviews on the computer, taking care to highlight illustrative verbatim quotations.[4]

Analysis and discussion

Perusal of the data suggested that different effects of the participatory evaluation project were apparent at different levels within the organization. Effects were observed at the levels of the evaluation team, the school in which the innovative pilot project was carried out, other schools in the district, and at the level of the central district office. Figure 6.1 shows the relative impact experienced by each of these groups, with the evaluation team being most affected and members of the other schools being least affected by the evaluation data.

Members directly and indirectly involved with the evaluation committee included the school principal and members of the pilot project steering committee who kept abreast of the committee's activities. At the school level, impact was considered in terms of the influence on members of the entire professional staff at the high school, including those directly involved in the implementation of the pilot project and those involved only in school-based decision-making about it. Members of other schools in the district, notably those having a stake in the pilot project by virtue of their offering intermediate and/or senior programs, comprised the third level of impact. Finally, members of central board office including senior administrators and members of the board of trustees comprised the fourth. It is important to recognize that, while the information in Figure 6.1 conveys a crude sense of the *degree* of impact of the study, the *nature* of that impact varied markedly. We now turn to an examination and discussion of the consequences of participatory evaluation at each of these levels.

Evaluation team

The impact of participating in the research process, particularly for members of the evaluation committee who were directly involved, was abundant. The process provided them with countless opportunities to interact with one another, often in very intense and substantively deep ways. The core team spent an

---

3 Further details about the study are provided by Cousins and Walker (1995).
4 Portions of these data are reported by Cousins (1996b).

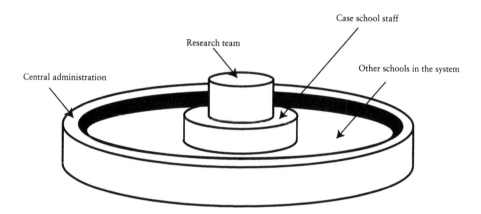

Figure 6.1. Relative impact of the evaluation by organization levels of analysis.

enormous amount of time and energy on the project over a protracted period of time. As the chair of the committee put it:

> It was fortunate to have a group of dedicated professional people willing to put in time and energy and the painfully long hours collaborating together, to interface with the data. There was an element of surprise that people with full-time teaching jobs could involve themselves in this ambitious program. They lived and breathed and ate it for a couple of years.

The intensity of the participation provided team members with opportunities to collectively refine their understanding of the pilot project, its fit within the school and how to improve it. It seems unlikely that superficial or "arm's length" participation would have led to the same result. As an example, I noted the committee's high level of engagement in a very technical discussion we had. The problem concerned a "glitch" in the quantitative data analysis of the teacher-importance ratings of reasons underlying their vote in the referendum. It became apparent to me that these team members were intimately connected with the data and had developed sophisticated understandings of the complexities they faced.

> This was a very interesting technical discussion that we engaged in basically at the level of lay-person discourse. I was impressed with the fact that they had observed this difference [in the statistical output] and identified this interpretation problem and it struck me that folks were very well acquainted with the data. (participant observation note)

Some of the follow-up interview responses provided confirmation of the development of deeper levels of understanding. "I learned a lot about leadership in implementing major change. It helped to analyze change in the school. [I became] more aware of difficulties within classrooms in implementing change. Personally, [there was] a tremendous amount of professional development" (steering committee member). "The staff involved in the research grew considerably as a result of it. It gave them the opportunity to deal with change issues, and to interrelate with people they normally would not have interrelated with on staff" (superintendent).

Part of the professional development experience extended beyond developing an understanding of organizational change and the pilot project into the research skills domain. Several of the research team members agreed that they had learned considerably from their involvement and that they would be willing to participate in future projects. However, a number of participants reflected on their relative naivete at the outset concerning their expectations for involvement. If they were to do it over, some very serious negotiating would be likely to take place prior to a decision to commit.

> Personally, I think there was stress . . . whether it was [school activities] I was involved with and the [nature of the] kids I was involved with or whether they were both part of the problem. Or that this [post-intervention] year we weren't doing the study and it was a much more enjoyable fall—less things, less stress because of [not having to do the study]. . . We had a much more pleasurable time because I was less stressed running around doing all sorts of crazy things. (evaluation committee member)

While the partnership with myself as someone able to provide technical research skills and resources was reassuring for the team, the addition to their workload proved to be extremely taxing. "There was a very heavy workload. People could not take the time to go outside the school . . . they did research on top of their regular jobs . . I think fatigue was a critical factor in a lot of people" (steering committee member). The fit of participatory evaluation within the existing workload framework is highlighted here as an extremely important consideration. While good fit has been something that we have advocated from the outset (Cousins & Earl, 1992), the potential for exacerbated down-side consequences is greatly enhanced where allowances for participation are not underwritten by administration. Weiss (1991) made the point several years ago and some of our recent data underscore it (Cousins & Earl, 1995). To address the problem in the future, the evaluation committee chair suggested that blocks of time would be worthy objectives for negotiation, because the mere provision of release time did not prove to be overly helpful: "People must appreciate that when a person takes time out of a classroom and gets a supply teacher, that sometimes it doesn't make the work less; it creates more work. It's often double work."

Since not everyone in the school—and indeed, beyond—was philosophically in agreement with the pilot project, the evaluation naturally became somewhat of a "political football." This was problematic for the evaluation team, since it created dissention and raised acute concerns about how the committee ought to posture itself within the school. On the other hand, such concerns provided the impetus for extended discussions among the team members about fundamental issues associated with the pilot. Ultimately, the evaluation committee took a neutral stance and drove themselves to report the data accurately and honestly knowing full well that, first, they were perceived by staff as being "pro-pilot" and, second, their reports of negative findings likely would be used as political fodder by the "nay-sayers." The political volatility of the exercise not only had implications for deliberations among team members but forced some to revisit their commitment to the project. One team member in particular stepped down from the team once negative data came to light. This individual had always raised concerns about the burden of the workload, a fact that may have at least partially explained his departure. Regardless, he rejoined the team in the final stages of the report production process after the committee had worked through the political issues and were well on their way to the final frank presentation of the data. The eleventh hour reappearance of the member in question caused me to wonder about the role of political considerations in his decision-making (participant observation note).

In summary, the impact of the participatory research process was quite significant for team members and those closely associated with the research committee's activities. Depth of participation and the political volatility of the focus for evaluation appear to have stimulated the development of their thinking about the innovation and its complexities as well as having forced them to visit deeply-held personal assumptions. In this sense, the evaluation acted as an interpretive learning system and double-loop learning was the result. The process, although professionally gratifying and rewarding, was also quite stressful, stemming chiefly from substantial increases in workload and unanticipated extended debates about the meaning of data and their relevance to the politically "hot" pilot project. Nevertheless, it was precisely these extended debates that precipitated deep levels of collective learning about the pilot initiative within the school context.

School staff as a whole

The impact of the research study on the school staff was viewed differently by different people. There was some evidence to show that the evaluation really helped staff to come to terms with the innovation and how it could work better. This was particularly evident in the school's response to the Ministry's provincial implementation of the transition years initiative one-and-a-half years after the pilot had been voted down by staff. Compared to other schools in the district—and indeed the province—the case school was viewed by interview respondents to be "well ahead of the game." While this is no doubt at least partly attributable to the virtues of having run the pilot, the evaluation cannot be divorced from that process. Indeed, the evaluation pointed out that too much attention may have been given to "second order" change factors while not enough attention was given to "first order", classroom-based changes. Staff streamlined their implementation efforts considerably from the initial pilot effort. According to the committee chair:

The greatest impact was on the school itself. Not having a global per-spective keeps them from appreciating the pilot. Interestingly enough, people at [the school] did not perceive that the study was impacting decisions, even though there were many data-driven deci-sions/recommendations made based on the findings. In particular, there have been radical changes to the pilot as a result of the data col-lected . . . They have embraced the change and have run with it. They are striking the 40–50% of the grade 9 teachers who are embracing transitions as an opportunity to impact the young people . . . Other schools are still struggling and some that haven't moved far beyond the few teachers who see it as a good idea and those who have no teachers wanting to teach grade 9.

Others tended to agree but they were guarded in their willingness to ascribe sig-nificant changes to the evaluation effort. As one of the research team members put it, "The research on the pilot project had a minimal impact on the school: it enlightened staff as to what colleagues were thinking and illuminated work-able strategies for transition years issues. But there was no follow-up to see which strategies carried on." The principal, on the other hand, was of the view that local, school-based impact of the research was noteworthy. "We certainly learned about the implementation dip—it was well noted and talked about." He had the sense that his staff was relatively happy with transition years imple-mentation issues compared to other schools in the system.

But not all of the informants were uniform in their assessment of the study's impact in the school. One of the evaluation team members suggested that the impact of the study was very much a function of prior dispositions.

[The evaluation team member] says that you would probably get 4 or 5 levels of impact if you separate the group that was immediately involved with [the study]; they would be most aware. Then there are a few more that are probably reasonably close to that group that would have a higher level background. The next level would be those who were possibly open to the material that was there. The next group would be people that were not open to any of that information. The fifth group would be the people that would use that information for negative purposes—those resistant to change. (interview fieldnote)

This perspective raises, once again, the role of political forces in determining the impact of the study. As it turns out there was some debate about just how political things got. The following comments highlight the discrepancies in opinion.

[The superintendent] felt that the study reinforced every person's opinions about what should have happened in the first place, or what was happening in the first place, and that the information and data did not tend to have the credibility that they once did. (follow-up interview fieldnote)

> Data and information is manipulated to support preconceived notions. People look at data and deny it or rationalize it or justify it or dismiss it." (superintendent)

> The political aspect tempered how people used the study. Those who wanted to fail [those implementing the pilot project] were holding us as disciples of the status quo. Neither party really read the study and understood it, in my opinion. That's an overgeneralization. I think a lot did, but the political nature of the study got in the way. (steering committee member)

> The report was not an avenue for supporting one point of view or another because people had to reconcile that this thing was happening and that an element of consultation had broken down with the Ministry because the Ministry was going ahead regardless of how they perceived the success or lack of success of the projects without a lot of input from the pilot projects as far as [the evaluation committee chair] could see. The exercise was designed to do that as painlessly as possible. They haven't quoted the report for one position or the other, but they have quoted the [referendum] vote. (interview fieldnote)

While there was some debate about the influence of political issues within the school, lack of uniformity in agreement about the utility and philosophy of the pilot project may be symptomatic of a more fundamental lack of consensus about valued goals for education. Given the nature of the pilot, probably at the heart of the discrepancy were debates about equity versus excellence as being the dominant driving force for schools. Forss, Cracknell and Samset (1994) report data suggesting that organizational learning is greater where dominant knowledge structures are well developed and have little diversity. In such situations involvement in evaluation will be one of the major instruments of learning. But, in the present case, a well-developed knowledge structure does not describe, in any way, staff's views about the innovative pilot or the goals it was designed to achieve. If we accept as plausible the argument put forth by Forss et al. (1994), the lack of goal consensus in the case school would act to limit rather dramatically the potential for use of the evaluation data and consequently for organizational learning. While some staff may have experienced relatively deep levels of learning, for others learning would be superficial. For still others, ideological dispositions may have precluded any learning from the evaluation.

Some relatively dissenting views about the participatory evaluation process itself were expressed. As one of the research team members said:

> The research report focused on administratively implementing a major innovation, rather than a "grass roots" reaction . . . When you document it [under an educational administration framework] you document it in terms of what control moves were made to make this work.

He went on at some length, making explicit his views about how the "alleged" participatory process had really been a sham. While no one else offered similar views and little evidence surfaced to support this team member's contention that mischievous use of the evaluation results were apparent, his sentiments raise questions about the fit of the participatory process in a politically turbulent context.

> The study leaned toward legitimatory research, a verification of what we did was alright, when it really wasn't. This was participatory in that we engaged local practitioners and we had a high-level man (Cousins) bringing us along . . . I think there's a game being played that's trying to make staff feel they're all participating in one nice, happy family. Wrong! "We're asking you the questions that we really want to ask and we're going to use the information basically to maintain what we want."

Flying in the face of these allegations are the data reported in the final document and the aforementioned commitment by the committee to reveal all. As mentioned above, the evaluation committee was composed of members who were unquestionably sympathetic to the pilot and its goals; but they were also an unlikely working group within the school community. The principal helped to put this in perspective:

> The make-up of the committee from a personality point of view was a concern right from the beginning—what [the committee chair] had was the leper colony. Because of this, the credibility was already under question before they even put pen to paper, but I think [the committee chair] did a hell of a job as the debate raged and evolved. There was a real evolution to [the evaluation process], I think it gained credibility. I think it started out as a sort of "what the hell is going on here?" kind of a committee to something that was treated a good deal more legitimately by most people.

He went on to describe how the negative findings concerning the effect of the pilot found their way into the school community through informal means as the study unfolded. "The committee was not leak-proof."

Another dimension that begs comment in the context of school-level consequences is the role played by me as the professional evaluator. While most were in agreement that the school–university partnership and its working relationship were fully functional, school staff expectations for the evaluation project may have been somewhat inflated as a consequence of my sustained involvement over such a lengthy period of time. In short, some staff may have developed the view that the final document would be much more definitive than it was.

> [A superintendent] believed that the study had a negative impact in the immediate educational community. The negative impact came from the amount of time and effort, the belief that the study was

> going to do a number of things and that these were probably false
> assumptions—that is, that the study and pilot were going to do more
> than they were capable of doing. (interview fieldnote)

Also, the document that was produced was comprehensive and although an
executive summary was widely circulated, some felt the final product was not
well enough tailored to the teacher audience. Some felt that workload demands
leave teachers with precious little time for reflection and therefore a document
of this sort would be viewed as overwhelming. Others, however, were of the
opinion that teachers' heightened curiosity and ownership of the pilot project
would more than offset these influences.

To summarize, the impact of the evaluation process at the school level was
relatively modest, although variation in opinion about impact was evident.
Some of this learning took the form of deeper understandings of transition years
initiatives and strategies and the identification of factors that inhibit goal attain-
ment. Lack of goal consensus and variation in philosophical subscription to the
pilot project and its purposes limited the perceived utility of the evaluation and
may have resulted in superficial or little learning for many. There were also indications that researcher involvement may have created false expectations and
placed limitations on the communication of findings to staff. On the other hand,
the apparent honesty and openness of a evaluation committee whose credibility was an issue from the outset may have enhanced the extent to which staff
took the study and its findings seriously.

Other schools in the system

The impact of the evaluation on other schools in the system was disappointing.
According to the research committee chair, various bits and pieces of the study
were picked up by these other schools as they began to grapple with transition
years implementation issues following directives from the Ministry. Some
schools ignored the findings altogether. For others, concerns about the events
leading to the case school's decision to hold the referendum and the ensuing divisive effects on staff overshadowed the potential usefulness of the evaluation
data. A superintendent disagreed, suggesting that it was the local feeder schools
that were most affected by the vote, and that since other area schools "didn't
have the same amount of ownership and personal experience attached to things
not continuing" they stood to benefit more directly from the data. Survey data
did not seem to bear this out, however. Written comment data suggest that the
vote and the negativity associated with it were prominent on people's minds. "I
don't know the impact [of the study] on teachers, other than the fact that a
majority of teachers voted to abandon the reorganized grade 9 program after
one year." "Word of mouth is entirely negative on the project, so I assume that
the study itself has had quite an impact." The impact these events had on those
directly involved in both the pilot and the research was significant.

> A point of discouragement to some people on the steering commit-
> tee of the project was that they were making somewhat of sacrifice

for the system, and yet the system ignored it and went ahead and made the same mistakes. It was disheartening and somewhat demoralizing for them. (evaluation committee chair)

The principal shared many of these concerns. He spoke of the "lack of success in profiting from the experiences in your own backyard" and "the failure of people, having seen the results in print, to internalize them sufficiently." His view was that many of the mistakes made during the pilot were being repeated. The research committee chair concurred that at least one principal dismissed outright the data as being non-credible and went ahead with a similar plan to the original pilot.

Survey data, however, were not uniformly supportive of this interpretation. A number of respondents revealed that other schools now faced with the implementation task had indeed profited from the case school's experience. "[The study] helped us to avoid [the case school's] mistakes and capitalize on successes." "It's given us ideas and shown us where and where not to go." "It was useful to help avoid the pitfalls." Others wrote about their deepened understanding of implementation issues and issues associated with the transition years initiative. "[I] personally became more aware of the implications of destreaming, both positive and negative." "The general aim of the study should be to rethink practices in a meaningful way—to that end [the study] was successful." "[It] increased insight." "Using results to change curriculum." This being said, it should be noted that the majority of survey respondents provided comments revealing that they were either unaware of the study's results or that the study had either no, very little, or a decidedly negative impact.

There were varying opinions about why this was the case. As mentioned above, the events and consequences associated with the referendum made some administrators a little "gun shy" about embracing the consensus-based democratic leadership model. This was evident, for example, in their reluctance to involve parents in decision-making processes associated with the grade 9 initiatives.

> "Other schools are very reluctant to involve parents. The fear is that they're going to get interest groups who will pressure decision-making into an area that they don't want to go. (evaluation committee chair)

By contrast, after having lived through the consequences of having the pilot voted down by staff, leadership at the case school persevered with their valuing of and efforts to seek input and partnership from parents. In the follow-up interview the principal indicated that he had announced to parents that the school wanted to form an advisory council for the transition years implementation made up of parents, teachers and students. He got 33 volunteers and subsequently invited these parents to a formative meeting, which—he was happy to report—was very well attended and largely successful.

Apart from fallout from the vote, one of the evaluation committee team members speculated that the approach to dissemination may have been limiting.

He thinks [the report] got buried, but he doesn't know who buried it. He made a point of asking people (superintendent, people at conferences, principals) if they were aware of this piece of material and only one person had heard about it. They distributed it and had a professional development day later on it and people asked about it and were told there was copy at their school. Maybe the five-page summary was not "colourful" enough, but he thinks it's a time factor with teachers with so much other curriculum stuff and it is not a priority. They keep starting a whole lot of things and there are not a lot of endpoints. (interview fieldnote)

Another research team member commented on how the dissemination responsibilities were essentially left to the principal and the research committee chair. He felt that these folks may have been perceived as having a vested interest in the results and this may have limited the credibility of the information being shared. A superintendent concurred:

There was a real enthusiasm out of [the evaluation team] that they wanted to share their findings with as many people as possible, which I viewed as being very positive. The down side is that perhaps one or two people took full ownership for delivering the message of the research and they didn't really allow the research team to be the deliverers of the message. Had that been my prerogative, I would have had more of the team share versus the leadership share, because the team did the work.

Further to the point, as external evaluator I was not directly involved in the dissemination process following the release of the final report. To be sure, there was some involvement on my behalf in sharing interim results and results from a provincial study with the case school staff and that these were received reasonably favourably (participant observation note). But other demands on my time limited my involvement late in the project. It seems likely that more direct participation in dissemination activities would have helped to raise the profile and perhaps the credibility of the report by offsetting suspicions about the motives of the principal and research team chair. Some evaluation theorists provide compelling arguments for a stronger role in follow-up for evaluators (e.g. Huberman & Cox, 1990).

Finally, survey data revealed mixed feelings about the utility of local applied research in general. While some respondents acknowledged this sort of research as being useful, practical, and relevant to local circumstances, others were much more guarded in their views and shared certain apprehensions about the potential for mischievous uses of local data. "No, it seems to be used to support a predetermined opinion or set of ideas." "Board level research is conducted by people who are virtually too far removed from the reality of the school environment; therefore, the research is generally not useful." "Not very [useful]. Often teachers do not get the true picture. Reports appear to be tailored to reflect what the Board wishes to reflect." ". . . teachers are never shown how research can affect their own classroom, or how ideas can be applied in the classroom." While these comments are not directly connected to the focal evaluation study, they raise questions about the extent to which applied research is integrated into the culture of the schools. Comments such as "teachers don't have time to reflect" suggest that much ground needs to be covered before applied social research activities truly become part of the organizational learning culture. A superintendent summarized the ideal rather nicely:

> The one thing you really come to know as an administrator is that there has to be a culture of learning established in the school environment so that change is not viewed as being dramatic, but that it's an ongoing process. If we are learning together, no one has to feel they are the target of the success or failure of a project. Everybody has a responsibility for their own learning and the contribution to the larger culture of learning in the school. You have to nurture, motivate and provide opportunities for that culture of learning, for risk-taking.

In sum, although some glimmers of impact of the case school's experience with the pilot were evident among staff in other schools, by and large this impact was fairly limited. From an organizational point of view, one would conclude that any learning that did occur was at a very low level. Fallout from events that unfolded during the pilot, limits on the credibility of data due to choices in dissemination strategies and the receptiveness of the system culture to the sorts of benefits that local applied research might have to offer appear to have limited the extent to which other schools were able to profit from the documented experiences of the case school.

Central administration

Compared with the non-case schools in the system, substantial impact on central administration seems to have occurred. The type of impact, however, was distinct from that encountered at other levels. To be sure, central administration profited from the substance of the report. In conjunction with other research being carried out province-wide on the transition years issues, these data, according to a superintendent, were used for setting directions within the board. "Out of that process we gained considerable insight into directions we needed to go in." Another superintendent reported using recommendations from the report as leverage for strengthening connections between the elementary and secondary panels and for integrating curriculum development thrusts into principals' and teachers' growth objectives. But the real impact of the participatory evaluation project on central administration appears to have been more of a level of support for the decision to continue with "home-grown" research as a means of generating, acquiring, and synthesizing local knowledge for decision-making.

The research committee chair commented on evaluation as a source of support for the creation of a system-wide research officer post, which he was recruited to fill.

> One trustee could say that she could now go to annual board of
> trustee meetings with her head held high that they [local staff] had
> done some original research and developed a quality document and
> that they could now make some political decisions based on data.

A superintendent concurred. She cast the newly created position as part of the
reorganization of program department,

> because the board saw a need to better demonstrate accountability
> to teachers, students and public. We used to do research conducted
> from the outside-in versus the inside-out. Now we have a person
> who can liaise with the outside world and also initiate inside-out
> types of things. Research was often an add-on to other people's jobs.

The research evaluation chair spoke of "a real climate of appreciation" concerning the research activities in which he had been engaged since his appointment to the new post. He had been "getting a lot of good feedback from senior
administrators" on the research and data that he had been able to generate by
himself over the past few months.

> The general sense is that, when teachers say things, they like that
> their responses are analyzed and that people do gather the data and
> make decisions based upon it. So they should take some time and be
> serious about the questionnaires and on the team-building exercises
> if it's going to be recorded and analyzed there is an appreciation for
> the research which is being developed.

Asked whether applied research was likely to continue in the board, a superintendent's response was summarized as follows.

> She sees research as definitely positive and she believes her colleagues
> around the administrative table would feel the same way, but she
> says it with caution. People would have to sense that there was some-
> thing worth promoting and that they would see direct payoffs.
> Unless the research was pinpointed and clearly defined with a con-
> tinuum of potential payoffs, then she thinks there will be great diffi-
> culty. They don't have an educational forum right now which would
> allow them to experiment as much as they had previously. Things
> must be concrete, applicable, immediate, and financially possible,
> then research would take on a different perspective and research
> wouldn't be dismissed. (interview fieldnote)

The foregoing suggests that while, on the one hand, central administration is
attracted to the prospect of continuing research and willing to support it financially, it would be only conditionally likely to do so. Two central purposes,
accountability to the taxpaying public, and program improvement are the most
likely "payoffs" for the Board. Both of these two central purposes for applied
research lend themselves to either legitimatory exercises or to incremental learn-

ing. While applied research of this sort holds the potential to raise questions about deeply-held assumptions and beliefs or double-loop learning, the extent to which it will foster change that would influence practice in fundamental ways is not yet clear.

In summary, senior administration also profited from the research process that unfolded at the case high school. Combined with other information circulating around the province, data emerging from the study helped administrators shape system-wide direction and provided them with leverage in some respects. More noticeably, the research exercise was used as part of a rationale to create a research position within the Board and to ensure the continuation of applied research activities for years to come. This administrative decision may be seen as one way to enhance the learning capacity of the organization, although at the time of data collection it was far too early to assess impact of this sort. The extent to which an organizational research mechanism can be used to foster deep levels of organizational learning remains to be determined.

Conclusion

This study examined the contribution to school district's collective learning and capacity development of activities associated with the collaborative evaluation of an innovative pilot project. Deliberately, the approach was one of providing fairly thick description of events and relationships among them and limited interpretation. Results reveal different patterns of influence on organizational learning at different levels within the organization. These effects were due in some cases to the participatory form of the evaluation activity, the central focus of the chapter, and in others to the substance of the evaluation results themselves. Clearly, these two sets of effects cannot be fully separated. But it is useful to explore the distinction in these concluding comments.

Most affected by the participatory nature of the evaluation experience were those directly involved—the evaluation team, and to a lesser extent, those who provided the team with their data. In the case of the evaluation team members, participation seems likely to have increased their individual research skills, and given them a deeper understanding of the change process (especially its politics). Such participation also increased their knowledge about colleagues' beliefs, and sensitized them to the amount of work entailed in projects such as this. The team as whole, no doubt, learned how to function well together, and such insight might be transferable to other groups in which they might participate.

At least some of those providing the evaluation team with data, and who also were involved directly in the implementation of destreaming, were stimulated to think more reflectively about their own practices related to destreaming. Their participation in providing data confronted them with questions about their practices which they may not have encountered in the absence of the project. And certainly the results of the evaluation seriously challenged some of these people to revisit their own commitments and beliefs about the innovation they were in the midst of implementing. While these staff members were able to rethink their approach to the innovation, and sometimes experienced substantial conceptual growth, political turmoil and lack of consensus on goals may have dampened their learning. Would the impact of the project have been greater in a less turbulent situation? This question is certainly worth pursuing, although high levels of turmoil seem now to be more the norm than the exception for schools. In this sense, the case study reported here should not be considered unusual or exceptional.

Those people more distant from the evaluation process varied enormously in the nature of what they learned from the project and the extent of such learning. Some staffs in other secondary schools in the district appeared to explicitly avoid learning from the results of the project. This seems to have been due to the informal communication network and the varying opinions flying back and forth about the case school staff's decision to shut down the pilot program for a year. These events overshadowed the extent to which others were willing to embrace the case school's experience as documented in the evaluation report and to vicariously learn through it. Nonetheless, some others were more open to the results and seemed to factor the results into their own plans for implementing destreaming.

The group outside the evaluation team that seemed to learn most from the project was the senior leadership team of the board. Motivation to learn was high, since they were responsible for ensuring successful implementation of this innovation throughout the district. They could not afford to ignore the results of the project. At the senior administration level, the most significant consequence was the development of the organizational infrastructure to enable future organizational learning to occur. The creation of the research officer post potentially enhanced the organization's capacity to learn, the direct results of which are unavailable for this study.

Based on this study, it seems reasonable to conclude that collaborative forms of evaluation of the sort described here do hold potential for fostering individual and collective learning. The act of collaboration, however, largely influences those directly involved in it, while the results of the collaboration may have wider influence. So, although it seems unlikely that any single applied research activity will lead to profound organizational consequences, the one reported in this chapter produced non-trivial learning of both an individual and collective nature. The effects of the strategy were leveraged, in this case, by the highly visible and political nature of the target innovative program, and the protracted period of time required to complete for the evaluation.

The likelihood of a novelty effect is present in this study. It will be interesting to examine other case studies where research units have been operating for lengthy periods of time and opportunities have been available for using collaborative evaluation, not just for legitimatory (accountability-oriented) or incremental learning (improvement-oriented) but also for more experimental and novel approaches to generating local, usable knowledge.

References

Ayers, T.D. (1987). Stakeholders as partners in evaluation: A stakeholder-collaborative approach. *Evaluation and Program Planning* 10: 263–271.

Argyris, C. (1993). *Knowledge for action: A guide to overcoming barriers to organizational change.* San Francisco, CA: Jossey-Bass.

Argyris, C. & Schön, D.A. (1978). *Organizational learning: A theory of action perspective.* Reading, MA: Addison-Wesley.

Bryk, A. (1983). *Stakeholder-based evaluation: New directions in program evaluation, No. 17.* San Francisco, CA: Jossey-Bass.

Cousins, J.B. (1996a). Understanding organizational learning for educational leadership and school reform. In K.A. Leithwood (ed.), *International handbook of leadership and administration* (pp. 575–640). The Netherlands: Kluwer.

Cousins, J.B. (1996b). Consequences of researcher involvement in participatory evaluation. *Studies in Educational Evaluation* 22 (1): 3–27.

Cousins, J.B. & Earl, L.M. (1992). The case for participatory evaluation. *Educational Evaluation and Policy Analysis* 14 (4): 397–418.

Cousins, J.B. & Earl, L.M. (eds.) (1995). *Participatory evaluation in education: Studies in evaluation use and organizational learning.* London: Falmer Press.

Cousins, J.B. & Walker, C.A. (1995). *Personal teacher efficacy as a predictor of teachers' attitudes toward applied educational research.* Paper presented at the annual meeting of the Canadian Association for the Study of Educational Administration, Montreal.

Daft, R.L. & Huber, G.P. (1987). How organizations learn: A communication framework. *Research in the Sociology of Organizations* 5: 1–36.

Fetterman, D. (1994). Empowerment evaluation. *Evaluation Practice* 15 (1): 1–15.

Fiol, C.M. & Lyles, M.A. (1985). Organizational learning. *Academy of Management Review* 10: 803–813.

Forss, K., Cracknell, B. & Samset, K. (1994). Can evaluation help an organization to learn? *Evaluation Review* 18 (5): 574–591.

Fullan, M.G. (1991). *The new meaning of educational change.* New York: Teachers College Press.

Huber, G.P. (1991). Organizational learning: The contributing processes and the literature. *Organization Science* 2 (1): 88–115.

Huberman, M. & Cox, P. (1990). Evaluation utilization: Building links between action and reflection. *Studies in Educational Evaluation* 16: 157–179.

Huberman, M. & Miles, M. (1984). *Innovation up close.* New York: Plenum.

Lant, T.K. & Mezias, S.J. (1992). An organizational learning model of convergence and reorientation. *Organization Science* 3 (1): 47–71.

Louis, K.S. (1994). Beyond bureaucracy: Rethinking how schools change. *School Effectiveness and School Improvement* 5 (1): 2–24.

Louis, K.S. & Dentler, R.A. (1988). Knowledge use and school improvement. *Curriculum Inquiry* 18 (1): 33–62.

Louis, K.S. & Simsek, H. (1991). *Paradigm shifts and organizations' learning: Some theoretical lessons for restructuring schools.* Paper presented at the annual meeting of the University Council for Educational Administration, Baltimore.

Lundberg, C.C. (1989). On organizational learning: Implications and opportunities for expanding organizational development. *Research in Organizational Change and Development* 3: 61–82.

Patton, M.Q. (1994). Developmental evaluation. *Evaluation Practice* 15 (3): 311–319.
Weiss, C.H. (1991). Reflections on 19th-century experience with knowledge diffusion. *Knowledge: Creation, Diffusion, Utilization* 13 (1): 5–16.

7

Professional Development Schools as Contexts for Teacher Learning and Leadership[1]

Linda Darling-Hammond, Velma Cobb, and Marcella Bullmaster

As other chapters in this book attest, the learning of individual staff members in schools is the foundation upon which organizational learning is built. It is by no means clear, however, that commonly used strategies for assisting such individual teacher learning accomplish their purposes. Many initiatives take little account of the teaching context; they do not view that context as a rich source of problems for stimulating learning. Often teacher development initiatives promote forms of knowledge not sufficiently grounded or sensitive to student and subject diversity to be of any practical guidance to teachers. Rarely do professional development initiatives draw on teachers' own experience, or redefine teachers' roles. And, typically, there is little attention given to the structural, administrative, and policy changes that would be needed for teachers to institutionalize significantly different forms of practice.

The last decade has witnessed a wide range of efforts to improve on typical strategies for fostering the individual and collective learning of teachers, as well as to better support, acknowledge, reward, or use teachers' abilities. These efforts range from from ladders and merit pay proposals that have aimed to identify outstanding teachers, to differentiated roles like mentor teachers and lead teachers, to the creation of professional networks and other learning com-

1 This chapter originally appeared in substantially the same form as "Rethinking teacher leadership through professional development schools", *Elementary School Journal* 96 (1): 87–106.

munities. Initiatives such as these variously hope to recruit and retain talented teachers, increase teachers' knowledge and skills, and motivate greater effort, more learning, or different practices on the part of teachers.

While seeking new roles or recognition for some teachers, many of these initiatives have maintained traditional views of most teachers' roles as implementers of curriculum decisions and procedures decided elsewhere in the bureaucracy. A small number of teachers are formally appointed to new leadership roles which, it is hoped, will help guide or support the others. These leaders are assigned new slots in the already highly-specialized administrative structure of schools. Although they engage in new kinds of decision-making, development, and mentoring, the knowledge, capacities, and authority of the vast majority of teachers are assumed not to change.

Other emerging strategies are more explicitly focused on redesigning the work, workplaces, roles, and responsibilities of all teachers—in short, redefining the job of teaching as one in which all teachers engage in decision-making, curriculum building, knowledge production, peer coaching, and continual redesign of teaching and schooling. Teachers in such settings assume roles traditionally reserved for "leaders." Their fuller professional role enables them to learn and lead continuously as they inquire together into ever more responsive practice. This professional conception of teaching relies on and promotes greater knowledge for teachers as the basis for responsible decision-making, and is thus related to teachers' pre-service and in-service learning opportunities as well as the kinds of tasks they engage in. Such redesigned work contexts bring especially significant challenges and possibilities for teacher learning and so are instructive for better understanding of how such learning can be supported.

In this chapter, we trace some of the possibilities for new forms of teacher learning and leadership that permeate teaching and are accessible to all teachers who engage the broader professional roles available in professional development schools (PDSs)—collaborations between schools and universities that have been created to support the learning of prospective and experienced teachers while simultaneously restructuring schools and schools of education. In the course of co-constructing learning environments where novices can learn from expert practitioners, the more highly-developed professional development schools allow veteran teachers to assume new roles as mentors, university adjuncts, school restructurers, and teacher leaders. They also allow school and university educators to engage jointly in research and rethinking of practice, thus creating an opportunity for the profession to expand its knowledge base (to engage in organizational learning writ large) by putting research into practice—and practice into research. And they socialize entering teachers to a new kind of professional teaching role, one grounded in collaboration, critical inquiry, and a conception of teacher as decision-maker and designer of practice (Darling-Hammond, 1994a). In these ways, professional development schools can help instruct teachers who assume new leadership roles and help to build a future teaching force that assumes leadership naturally as part of a more professional conception of teaching work.

We examine the potentials of PDSs for fostering more widespread individual and collective teacher learning using data from in-depth case studies of seven PDSs that are among the more mature of these new institutions (Darling-Hammond, 1994b), supplemented by research in a number of other professional development schools where similar patterns of teacher leadership have been noted (Boles & Troen, 1994; Kerchner, 1993; McCarthey & Peterson, 1993; Teitel, 1992). The case studies, sponsored by the National Center for Restructuring Education, Schools, and Teaching (NCREST), were conducted using a common research design[2] in sites that had been engaged in PDS work for a number of years. Most of the sites are places where the partnerships between schools and universities were relatively longstanding and where the partners held joint intentions for simultaneous rethinking of both teaching and teacher education, although sometimes these intentions developed out of the collaborative process rather than preceding it.

Thus, these PDSs are more likely to represent the aspirations of PDS proponents (see, for example, Darling-Hammond, 1994a) than to be "typical" of the vast range of experiments that have been launched under that name over the last ten years. Among the more than 200 professional development schools that now exist, few represent all of the possibilities of PDSs and most are at various stages of struggling to invent collaborative relationships where none existed before (Darling-Hammond, 1994a; Duffy, 1994). Their infant efforts are highly varied. Though they have in common a striving to improve education for students and for current and prospective teachers, no single fledgling professional development school encompasses the entirety of goals for professional development schools writ large. Some emphasize redesigning pre-service teacher education; others have created innovative models for in-service development through restructuring and action research; still others have designed new teaching roles which renew veteran teachers and teacher educators.

Regardless of a particular PDS's starting point, a number of recent accounts suggest that the work of PDS development encourages beginning and veteran teachers, teacher educators and administrators to redefine their work and their working relationships in a variety of unanticipated ways. Our case studies suggest that, in places where the PDS concept has been growing for a number of years, teachers have begun to transform teaching fundamentally so that it incorporates leadership roles—defined in terms of functions rather than titles—as the norm for all teachers.

In our analysis of teacher leadership in professional development schools we make three major claims: that teacher leadership is inextricably connected to individual and collective teacher learning; that teacher leadership can be embedded in tasks and roles that do not create artificial, imposed, formal hierarchies and positions—and that such approaches may lead to greater profession-wide

---

2 The case studies were based on extensive interviews, observation, and review of PDS documents, surveys, and teachers' logs over a year-long period. Most of the fieldwork was done during 1991 and 1992.

leadership as the "normal" role of teacher is expanded; and that the stimulation of such leadership and learning is likely to improve the capacity of schools to respond to the needs of students.

In the course of developing these ideas, there are a number of things this article does not do. While we point to teachers' views that their practice has become more effective, the case studies did not assemble evidence that teachers' engagement in new roles yields greater learning for students. In addition, we do not treat in any detail the many problematics of establishing professional development schools and of creating new roles and relationships. These are treated elsewhere (Darling-Hammond, 1994a, 1994b; Duffy, 1994). Finally, while we recognize the importance of and changes in the roles played by principals, parents, and non-teaching staff in PDSs, a full examination of these issues is beyond the scope of this paper.

Teacher learning and teacher leadership in professional development schools

Professional development schools are a special case of school restructuring, aimed both at the creation of learner-centered practice within individual schools and at the creation of a profession of teaching comprised of knowledgeable, empowered teachers. As they have been conceived by the Holmes Group (1986, 1990), the National Network for Educational Renewal (Goodlad, 1990), and by school- and university-based educators involved in the development of some of the earliest models, professional development schools have a distinct mission: they aim ultimately to prepare all teachers to teach all children for understanding; to meet the diverse needs of whole children and families; to enact shared governance within the school community and in the relations between schools and universities; to redesign schools and schools of education for constructivist, personalized, and collegial learning; and to function as communities of learners. They are not intended to be merely laboratory schools or pilot projects: their aim is to transform the entire educational enterprise by changing teaching, schooling, and teacher education simultaneously.

Thus, in the eyes of many who are seeking to create these new institutions, the professional development school is a strategy that can potentially connect all of the elements of educational reform (Murray, 1993) and solve the chicken–egg dilemma of preparing teachers for schools that do not yet widely exist. The dilemma is that, if the goals of school reform are to be realized, teachers must be prepared to function in schools that are restructured to meet the needs of diverse learners well. Yet such schools cannot exist in large numbers until more teachers are differently prepared and the practice of veteran teachers has changed significantly. The resolution of the dilemma relies on colleges working in partnership with local schools to agree to change jointly and simultaneously both the manner in which teachers are prepared and the conditions under which they practice. In so doing, PDSs provide a mechanism that cuts across institutional boundaries to create greater knowledge and capacity across the entire profession. They are potentially powerful strategies for organizational learning, with the "organization" defined broadly to include school systems as a whole.

In this mission, PDSs challenge traditional ideas about who are learners and who are leaders within schools and in school–university partnerships. In the most highly-developed PDS sites, teachers work in teams with each other, with prospective teachers, and with teacher educators, discussing learning and learners from many vantage points; they examine the effects of their practice; they adapt practices based on evolving understandings of learning and learners; and they continually rethink school structures and teaching strategies. Veteran teachers engage in mentoring new teachers and co-constructing teacher preparation programs; both novices and veterans develop curriculum and make decisions about school and classroom practices; teachers lead problem-solving endeavors within and beyond school boundaries and participate in research within and beyond their classroom walls. The focus of PDS work on reconceptualizing teaching and learning creates new forms of teacher leadership linked to new forms of teacher learning.

In highly-developed PDSs, like some other restructured schools, these opportunities for leadership are available to all teachers, without regard to formal roles and titles. Perhaps more significant, professional development schools are preparing incoming teachers to see curriculum building, decision-making, school change, and research as part of their normal teaching role, thereby developing a new generation of teachers for whom leadership is a starting-point, not an end goal to be achieved at the close of a teaching career. In short, the conception of teaching and the role of teacher are being redefined to include those responsibilities for developing knowledge and transforming practice that were allocated to other organizational slots when teaching was bureaucratized and de-skilled after the turn of the last century.

This conception of teacher leadership stands in contrast to the traditional, officially defined, pre-structured, "add-on" leadership positions portrayed by Smylie and Denny (1990) as the "individual appoint, anoint, and training" approach. Such positions have been problematic in at least two ways. First, they often violate the strong egalitarian ethic among public school teachers (Boles & Troen, 1994). Within the standard system, attempts to assign formal leadership roles to teachers often place would-be teacher leaders in direct opposition to their colleagues. Acknowledged differences in status based on knowledge, skill, and initiative are seen as taboo within the culture of teaching (Boles & Troen, 1994; Little, 1988) as they blur lines between management and labor and create differences among laborers where solidarity is needed (Tyack, 1974). Given this ethos, Little (1988) observes, "Teachers placed in positions that bear the titles and resources of leadership display a caution toward their colleagues that is both poignant and eminently sensible (p. 84)."

Second, recent research suggests that defining and assigning formal leadership roles to teachers within the conventional structure of schools may have some benefit for the teacher leaders themselves, but there is usually little additional learning for the "non-leaders." Smylie's (1994) review of research on teacher work redesign, including formal leadership roles, found that teachers who assume redesigned roles are more likely to learn and change their classroom practices than are the other teachers who are presumed to benefit from their work. Because the designated leaders (mentors, curriculum developers, and the like) have more opportunities for learning and for collegial interaction in their new roles, they become more professionally engaged and knowledgeable. In other words, teacher learning and teacher leadership are inseparably fused.

We highlight teacher leadership roles that have materialized in highly-developed PDSs where teachers serve as mentors and teacher educators, curriculum developers and decision makers, problem solvers and change agents, and researchers engaged in knowledge-building. We emphasize that these are roles that are being developed for all teachers, not just a few that are specially anointed or those who happen to work in professional development schools today, and that they influence new teachers as well as veterans. The role of the teacher embedded in these schools is that of an individual who transforms a knowledge base, reflects on practice, and generates new knowledge. The teacher must be a learner in order to teach, and, in so doing, the teacher comes to own and produce knowledge rather than being controlled by it. With this liberating process, the teacher becomes a leader as well—developing the knowledge and decisions that shape practice.

Teacher learning in professional development schools

The case studies we examined illustrate how professional development schools are creating new possibilities for teacher learning, as novices and veterans, school-based and university-based teacher educators have opportunities to learn by *teaching*, by *engaging in restructuring*, and by *collaborating* (Darling-Hammond, 1994a). These schools are developing around a constructivist understanding of learning for both teachers and students, one that acknowledges that as all members of the school community engage in formulating, testing, and enacting ideas, they forge new knowledge and create more profound understandings. The generative iterations of this cycle of teaching and learning locate control of the learning with the learners themselves, thereby involving them in leading rather than implementing both personal and institutional transformations.

Learning by teaching

Traditional frames for teacher education, like those for elementary and secondary students, have envisioned teaching primarily as information transmittal, and learning as the acquisition of knowledge by individuals. Whether, in courses or student teaching, clear distinctions between teacher and learner, expert and novice, have been maintained, along with tidy compartments separately housing theory and practice, knowledge and application.

In professional development schools, these distinctions begin to disappear. Teacher educators learn more about teaching as they teach collaboratively with veteran teachers. Veteran teachers find themselves learning more about both the theory and practice of teaching as they teach novices. The old saw that you really learn something when you teach it to someone else has proven true for PDS participants across many sites. As one teacher at Wells Junior High put it, "[W]atching somebody else teach and thinking how I might change it . . . forced me to really think about teaching . . . in a different way." Another echoed, "it helps you reflect on what you're doing as well as reflect on what the interns do" (Miller & Silvernail, 1994, p. 42).

While insights into teaching frequently occur when veteran teachers work with student teachers, the power of this reflection appears to be substantially augmented when a faculty collaboratively engages in developing a learning environment for a cohort of student teachers or interns. In a study of three PDS sites, Teitel (1992) found that the "deep engagement" of PDS teachers in the preparation of pre-service teachers caused them to report having very different experiences with student teachers than those in non-PDS environments, experiences that they feel benefit their own professional development and their enthusiasm about teacher education.

Their collective engagement in the process of preparing new teachers triggers deeper insights and the development of shared norms that are absent in the traditional idiosyncratic placements of student teachers to lone cooperating teachers. The case studies reviewed here suggest that, as classroom teachers become teacher educators, they find their own knowledge base deepening and their practice becoming both more thoughtful and more shared.

Learning by engaging in school redesign

In restructuring professional development schools, teachers also find that they are learning by actually engaging in the work of thinking through, researching, debating, and implementing innovations. Whitford (1994) describes how teachers at Fairdale High in Louisville, working in a PDS with the University of Louisville, were propelled into professional readings and conversations to answer the questions that emerged as they planned and attempted their own reforms. In this case as others, "much professional development occurred in a learning-by-doing approach" (p. 86). In such schools, experts and novices learn together about teaching—in newly forged intersections between research, theory, and application—as they do the work of school restructuring together.

Learning by collaborating

Perhaps one of the most promising aspects of professional development schools is that they emphasize collaborative planning, teaching, and decision-making within and across institutions in ways that redefine both the act of teaching and the nature of their home institutions. Most PDSs have introduced or strengthened existing arrangements for team teaching at the school sites and, frequently, for teacher education courses as well (AACTE, 1992). Beginners and veteran teachers learn from each other as they engage in cooperative team planning and teaching. Cohorts of student teachers or interns, when they are organized in pairs or teams, have many opportunities to collaborate with each other as well as with the teaching teams they work with at school.

Shared decision-making creates still other "teachable moments" within and across the several role groups of PDS inventors and participants. At P.S. 87, a PDS working in collaboration with Teachers College, Columbia, in New York City, one veteran teacher echoed the sentiments of many others throughout the professional development schools who found themselves inspired by new opportunities to collaborate with colleagues:

> The support I received through this ongoing informal and formal sharing was what gave me the courage to try something new. I never would have done it alone. Schools need to change so that we all have more contact with each other because teachers are out there alone. Ideas need to be passed amongst teachers just like they need to be passed amongst children. (Darling-Hammond, 1994a, p. 12)

Other research has found that teachers are most likely to engage students in cooperative learning experiences when they themselves have been involved in such opportunities. At I.S. 44, another New York City PDS working with Teachers College, the benefits of collegial learning extend from school and university faculty to student teachers and students. In a month-long school-wide project, children enjoy the benefits of collaborative learning as they work in groups on their interdisiplinary projects that are planned and supported by teams of student teachers, school-based faculty, and university-based faculty. In this hands-on experience of curriculum invention and interdisciplinary team teaching, teaching and learning roles become fluid and inseparable. Learning by teaching, learning by doing, and learning by collaborating occur simultaneously within both the teaching teams and the student groups. As Lythcott and Schwartz (1994) note:

> This was not a program in which "experts" mentored apprentices; rather everyone was learning together . . Released from a self-conscious mentor role, cooperating teachers began to focus on what they themselves had been able to achieve in doing things differently, and to build on those. Rather than explicitly focusing on what student teachers were learning about teaching, they assumed that that was going on and they simply talked, analyzed, invented, modified, and shared in conversation with them as one set of colleagues to another—a new model for student teaching. (pp. 136–137)

As a consequence of this experience, student teachers and veteran teachers acquired new frames for thinking about their teaching—frames that include professional collaboration and collegial problem-solving, interdisciplinarity, and "whole child" perspectives as foundations upon which to build their future learning and experience. They described themselves as "collaborating constantly," and able to understand more about their students by seeing them in other classrooms and subjects. A transformation of thinking about both students and subjects occurred as a result:

> Originally [each of us four student teachers] thought of ourselves as "experts" only in one insular discipline. Now at the conclusion of this program it would seem awkward for me to address a student as just a reader of English Literature and not as the rounded individual he/she is . . . The student and the discipline would suffer from this artificial narrowness. (Darling-Hammond, 1994a, p. 13)

These kinds of experiences for new teachers can create a different frame from which they will learn throughout their professional lives. If they see learning as continuous, collegial, integrated, and child-centered—and if they see collaboration and reflection as opportunities for learning—they are more likely to build new kinds of knowledge for practice and to use knowledge differently.

Teacher leadership roles

These opportunities for new kinds of learning become the basis for new forms of leadership for teachers. In PDSs like those we have examined, teachers' beliefs, experiences, personal knowledge, and values are all acknowledged as fundamental in guiding the active process of creative inquiry, analysis, and evaluation that are essential to the practice of professional teaching. These activities stimulate hands-on learning that in turn leads to an inventive approach to practice that "bubbles up" into new forms of professional, collegial leadership (Boles & Troen, 1994, p. 9) At the Learning/Teaching Collaborative, a teacher-initiated professional development school collaboration between six schools in Brookline and Boston, Massachusetts, and Wheelock and Simmons Colleges, teachers have changed their roles and the conditions of teaching and learning as they have formed teams, taken risks with their teaching, expanded their knowledge base, and relinquished their individual control to the collective judgment of the group. Of these teachers, Boles and Troen write:

> They are demonstrating their strength in ways they would never have imagined just a few years ago. The PDS has broadened their horizons beyond the school and exposed them in new and meaningful ways to the world of theory. They have seen their practice reflected back to them through the interns' eyes. As they assume new leadership roles they have deepened their understanding of policy, curriculum, and the value of research to practice. (p. 25)

Leadership looks very different from traditional bureaucratic, hierarchical conceptions that slot individuals into different, limited functions and that place them in superordinate and subordinate relation to one another. Rather than being defined by formal roles or positions, the leadership that emerges in these settings is more like Sergiovanni's (1987) concept of "cultural leadership"—the "power to accomplish" as opposed to "power over people or events." Leadership is widely diffused and flows from matches that evolve between teachers' expertise and interests with the inventive work that needs to be done, rather than consolidated into a particular position or role that has predefined functions. In PDSs, leadership emerges in organic ways that resemble Howey's (1988) notion of career lattices:

> The imagery of career lattices rather than career ladders would capture the dynamic interchange of roles and responsibilities envisioned. Leadership would tend to be differentiated according to purpose and need at the school site and attend in a highly specialized way to such functions as curriculum redesign and articulation, organizational monitoring, pedagogical development, or collaborative action research. (p. 30)

In this more fluid context, which acknowledges and uses expertise in many ways, teacher knowledge and leadership are recognized as a major resource to the school community, one that has previously gone untapped. A broader conceptualization of "collaborative leadership" enables principals, teachers, and teacher educators to work together in the design of organizational structures that build on their collective knowledge and commitments.

The Learning/Teaching Collaborative (L/TC), for example, is governed by a Steering Committee composed mainly of teachers with representatives from college faculty, and college and school administration. Every teacher participating in the collaborative sits on one of five governing subcommittees. These subcommittees enable teachers to take lead roles in everything, from governing the Collaborative's budget, to the organization of professional development activities based on teacher identified needs, to interviewing and selecting prospective interns in collaboration with college faculty, who will be assigned to L/TC PDS sites. As compared to traditional bureaucratic structures, Boles and Troen (1994) note that:

> The Learning/Teaching Collaborative, with its collective form of leadership assumed by many individuals, looks very different. In this leadership paradigm, teachers develop expertise according to their individual interests. They continue to feel professionally independent, yet they are part of a working team. No teacher has higher professional status within the Collaborative and a range of roles in leadership is available to all the teachers. Thus the role of teacher leader is reconfigured to be inclusive, rather than exclusive, and is available to significantly more teachers. (p. 19)

Similarly, at Lark Creek Middle School in Washington state, a PDS working with the University of Washington and the Puget Sound Professional Development Center, teacher leadership is enabled by governance and structural changes that use time, expertise, and human resources differently (Grossman, 1994). These include the formation of an instructional council, responsible for site-based decision-making; the reorganization of the staff into interdisciplinary teams; and the placement of all of the teachers and student teachers on one of six teams, with each team responsible for teaching approximately 100 students.

These instructional teams have become increasingly involved in making decisions regarding their work, including the scheduling of student classes, deciding budgetary matters, identifying problems facing the school, and proposing solutions. A site committee, composed of teachers from Lark Creek and a professor and graduate student from the university, makes decisions regarding the implementation of Professional Development Center projects at the school, the use of the budget for pre-service and continuing professional development activities, and recommends curriculum-related changes to the Steering Committee. Teachers take on responsibilities once formally reserved for others in the administrative hierarchy in a variety of ways as part of their expanding roles. Their decisions are based upon what is needed to support teaching and learning for students and teachers.

Through information gleaned from case studies of highly-developed PDSs, it is becoming clear that a form of teacher leadership is being created that holds the promise of being more than a set of formal, bounded, titled, and assigned roles within these schools. In these PDSs, teacher leadership is potentially more than a role; it is a *stance*, a mindset, a way of being, acting, and thinking as a learner within a community of learners, and as a *professional* teacher. Rather than officially slotting teacher leaders into predesigned functions, a process that requires the teacher to fit the designated leadership mold and the school to conform to the organization chart, PDSs offer the possibility of the school's taking shape around its team as the teachers contribute their individual interests, abilities, and experience to the community of learners and leaders.

Teachers as mentors and teacher educators

PDSs provide clinical sites for preparing beginning teachers in ways that integrate their teacher education experience with a structured internship under the guidance of an expert teaching faculty that is both school-based and university-based. Because this joint faculty forges a collective vision and a common program for their work and conducts the work as a team rather than as separate components of a fragmented program, teachers at well-functioning PDS sites can become full partners in the preparation of the next generation of teachers. In the collaborations between school-based and university-based PDS faculty, all partners can imagine and invent new images for teaching and teacher development.

Because effective teaching is context-based and must be adaptive to individual students, successful teacher preparation must involve not only a foundation of theoretical knowledge, but also a rich array of clinical experiences that help teacher candidates integrate their formal knowledge of teaching and learning with the knowledge of adaptive practice that can be gained only by working with the guidance of experienced teachers. As they create this "rub between theory and practice" (Miller & Silvernail, 1994), professional development schools have found themselves creating new, hybrid ways of knowing and forms of knowledge that have a special power and energy of their own: knowing through action and reflection as well as by understanding and appreciating the findings of others; knowing through sharing different experiences within a collegial group of practitioners; knowing through research conducted by teachers along with researchers that is informed by the diverse experiences of individual children as well as the aggregated outcomes codified in larger-scale empirical studies (Darling-Hammond, 1994a).

In the process, teachers take on the roles of knowledge producers and knowledge shapers as well as knowledge users. As one Teachers College, Columbia University, professor described the co-construction of knowledge among PDS participants:

> I think that the knowledge that they [teachers] have and the knowledge that they can construct with the students [student teachers] is part of the knowledge base of teacher education . . . I think what we get in interacting with experienced practitioners are new ways to look at what we're looking at through a research perspective. This is not only useful in terms of our work with prospective teachers, but useful in terms of the way we do our research: the way we interpret our findings; the way we conceive our research questions; what we think are the important agendas to pursue. (Snyder, 1994, p. 122)

At the Wells Junior High School, teachers' knowledge is also finding voice in new ways. Miller and Silvernail (1994) suggest that "what is happening at Wells are a series of private epiphanies about conceptions of knowledge and the appropriate role of schools and university faculty for sharing that knowledge." The synergy not only legitimizes teachers' knowledge, it enables teachers to use it in powerful ways beyond their individual classrooms.

> Integration of theory and practice goes beyond combining academic coursework and field experiences. In general, teachers' voices have been uninvited, unheard, and devalued in professional discourse about teacher education. The Wells PDS makes teacher voice central to its preservice program and acknowledges the unique perspectives, insights, and wisdom that practicing teachers have accumulated and incorporates these into the preservice program. Through ongoing, daily discussion, story telling, and reflective interaction, experienced teachers talk about the tacit understandings and informal rules of practice that underpin their knowledge of the teaching craft. Through continuous conversation in the context of real schools and classrooms, teacher voice assumes a privileged authority and often challenges the more formal knowledge base that university professors represent. (pp. 38–39)

As teachers assume leadership in co-constructing knowledge for teacher education, they also create more powerful learning cultures within their schools. At the Pontiac Elementary School, a PDS associated with the University of South Carolina, teachers are equal partners in structuring the experiences of student teachers and interns. For example, education methods courses are jointly designed by school and university faculty and are offered at Pontiac. Berry and Catoe (1994) found that the climate of the school is also changing as teachers' involvement in the education of future educators grows. As teachers assume leadership in connecting many of the PDS experiences to the school's efforts to revise its curriculum, the reforms:

> are indeed leveraging a learning culture for teachers, administrators, and interns. As a result of their PDS and restructuring efforts, over 70% of the teachers reported that they changed the way they reflect on practice while 61% reported they changed their conception of collegial work. Similarly, over one-half (55%) of the teachers reported that they have changed the way they teach and their conception of what needs to be known in order to teach. (Berry & Catoe, 1994, p. 184)

The same outcomes have occurred as teams of cooperating teachers work with student teachers at Lark Creek Middle School (Grossman, 1994). Members of the teaching team meet monthly with university site-supervisors from the teacher education program "to discuss how students are doing in the field, to plan aspects of the seminar or field experiences, and to share information related to supervision and evaluation of preservice teachers" (p. 57). Grossman describes teachers' views of the results of this process:

> Teachers talked both of the improvement in the preparation of future teachers represented by the PSPDC pilot program and of the rewarding aspects of their own involvement. Most teachers felt that in comparison with their own professional preparations, the student teachers at Lark Creek were receiving "vastly improved" preparation . . . Working closely with student teachers has helped teachers expand their sense of a teacher's professional role. "Teachers really do have a responsibility to the profession. That's part of what we need to do. Find the best student teachers we can to replace us," said one cooperating teacher. Teachers also felt that they were learning from the student teachers: "I've learned from student teachers different ways to teach things. They're teaching me too," commented a veteran teacher. (p. 63)

As these comments suggest, the process of teaching new teachers invites more learning, causing the processes of leading and learning to become deeply intertwined and interdependent. At Norwood elementary in Los Angeles, a professional practice school in collaboration with the University of Southern California school of education, university, and school faculty created an additional avenue for student teacher learning by developing a series of problem-solving clinics for interns, based on common concerns and problems. This, along with other responsibilities of the project teachers, was another means for taking leadership that stimulated a great deal of learning and development for veterans as well as for their interns:

> Involvement with teacher education and the study of how student teachers learn to teach has made a significant impact on the teachers' own professional practice at Norwood. We have reviewed models of teaching and have seen how to implement them. We have become aware of the value of peer coaching and the need to foster collegial relationships . . . The [problem-solving] clinics have had a definite impact on the project teachers, as some of us have helped write cases and lead seminars. This puts a fair amount of pressure on in-service professionals to keep current on educational issues and help identify causes of, and remedies for, poor student achievement and other classroom concerns. (Lemlech et al., pp. 163–164)

As teachers become mentors and teacher educators, as they assume greater responsibility for the collective profession, they also become more comfortable with the notion that seeking and leading collective improvements in practice are aspects of a professional role. In highly-developed PDSs, all teachers in the school participate in establishing a climate and a culture for the preparation of new teachers, taking on a variety of roles in doing so: modeling, advising, coaching, holding seminars, offering assistance. As Berry & Catoe (1994) observe of the steady march toward professional leadership in the Pontiac/University of South Caroline PDS:

> Few PDS planning sites view teacher education as their responsibility with the intensity that is found at Pontiac. But those Pontiac teachers who are most involved are distinctly different. These teachers are learning to recognize that they need to help enforce standards of practice. As one teacher asserted: "We want to be involved in teacher education, that's what the bottom line is . . . I feel like it is so important to us to send out good teachers . . . If we are going to get the respect that is due our profession, we have got to take charge and build our profession up. " Not only are more Pontiac teachers interested in finding time to see others teach, they are more interested in having someone . . . see them teach. And this openness can indeed transform a learning culture and change a school. (p. 187)

Teachers as curriculum developers and decision makers

Professional development schools engaged in rethinking curriculum, teaching, and learning create leadership roles for teachers as they redefine teaching from a formulaic exercise in which teachers implement curriculum designed by others to a creative act that is responsive to students' experiences, talents, and needs. In schools that are restructuring to take account of new understandings about how students actually learn, students are no longer seen as raw materials or empty vessels, teachers are not merely conduits, and the process of teaching and learning is not conceived as one of simply pouring knowledge into students' heads. Rather, knowledge is viewed as interactively constructed by learners and teachers who function in a reciprocal relationship within a community of learners. Students' learning is built upon their prior knowledge, experiences, and beliefs, with the teacher creating bridges between these very different starting points and common curriculum goals. Thus, curriculum is always in development and adaptive teaching strategies are always being discussed and invented. A constructivist approach to teaching and learning means that teachers are curriculum developers and assessors who continually make decisions about varied content and methods.

The case studies illustrate how teacher-leaders in PDS sites that are striving to develop and model this dynamic, interactive view of teaching are building and using content knowledge and pedagogical knowledge in ways that support the growth and development of diverse learners with multiple learning styles, intelligences, family and cultural backgrounds, and life experiences. They develop curriculum with and for their students and the prospective teachers with whom they work. In connection with this reshaping of curriculum and teaching, the question of assessment of student learning also comes to the fore, since teacher and student learning are inextricably connected. Richer forms of assessment that enable teachers to understand how their students learn as well as what they know enable teachers to be more effective. Thus, as teachers develop ways to learn more about their students' thinking and performance, they also become assessment developers, observers, and documenters of learning. They talk about students' work and discuss what "good performance" means and how to support its development. They take leadership in developing curriculum and assessments and in beginning to define shared standards of practice. New teachers learn how to become professionals whose scope of expertise and decision-making in these spheres is broader than it has ever been before. They grow into leadership roles from the very beginning of their careers.

The stimulus to invention is powerful. Snyder (1994), for example, found that the professional growth of experienced teachers at P.S. 87 and I.S. 44 in New York was stimulated by creating formats for collaboration, attesting to the fact that "all collaborating educators, no matter their depth of experience or whether school or university based, reported creating totally new curriculum units—some for the first time in years" (p. 107). Teachers at the Learning/Teaching Collaborative in Massachusetts develop curriculum together, seeking to "reimagine curriculum" and "reconfigure the way kids learn," with the resources of the college, the assistance of their interns and their peers, and the Alternative Professional Time that is allocated to them during the school day (Boles & Troen, 1994). At Lark Creek Middle School collaborative curriculum development has become a means for developing collective understandings and practices. Grossman (1994) notes that "perhaps what is most striking about Lark Creek is the widespread use of a common language with which members talked about the activities of teaching, learning, and the process of reform" (p. 70):

> I don't know of another school that talks as much about practice,"
> explained [a] teacher . . . A visit to Lark Creek reinforces the image

of a collegial school . . . Visitors might find teams of teachers working together to solve school problems, or walk in on a science teacher
conferring with a science teacher from another PSPDC site about a
new science curriculum. (p. 61)

At the Norwood–USC professional practice school in Los Angeles, teachers
moved from tentatively engaging curriculum issues to taking initiative with
respect to student assessment and staff development. Lemlech and colleagues
(1994) describe the evolution:

> As the PPS teachers discussed their instructional goals at the initial
> goal-setting session, we were reassured that both school and univer-
> sity shared many of the same beliefs. Through much discussion, the
> planned and evolving instructional program centered on the develop-
> ment of a thinking curriculum organized around thematic units that
> would integrate disciplines and use a variety of instructional strate-
> gies. Project teachers began to work in partner teams to plan the-
> matic units after first requesting help from the university participants
> on how to develop teaching units. (p. 161)

> As the project progressed, individual members assumed leadership
> to discuss the integration of assessment with teaching and learning.
> At a recent all-day meeting, PPS teachers led the session focused on
> authentic assessment. They prepared exhibits and engaged PPS mem-
> bers in discussion of ways to have children demonstrate what they
> are learning. This served as an example of teachers taking charge of
> staff development time, making key decisions about what they want
> to study. (p. 165)

Kerchner (1993) found that at Byck Elementary School, a PDS associated with
the Gheens Academy in Louisville, Kentucky, the participative management
structure has given rise to a score of teacher initiatives in the classrooms, including the creation of a four-teacher, cross-age team to encourage students from
kindergarten to grade 5 to work and learn together, and interdisciplinary teams
at the third- and fourth-grade levels. Also, two first-grade teachers and their assistants at Byck knocked down the wall between their classrooms, creating an open-
space learning environment for small homogeneous groups. Kerchner observes:

> None of these programs is unique. But each is a local invention—a
> "little try"—that gives teachers ownership of an academic enterprise.
> Not all the programs follow the same pedagogical assumptions . . .
> The common thread among the programs is that they seem to elicit
> high levels of commitment from the teachers who originated and
> operate them. (p. 36)

McCarthey and Peterson (1993) show how this process of gradual assumption
of professional initiative unfolds in their case study of two elementary teachers
in a restructured professional development school. One of the teachers, Julie

Brandt, who was a designated team leader of four teachers and 120 students, took on leadership of a process of collective change as she was herself supported by the professional networks surrounding the PDS and by the shared decision-making process adopted by the PDS:

> Brandt attributed the changes in her classroom practice to the restructuring efforts, her work with . . . the liason from the professional development center, and her other connections at the center. Brandt went to the center for materials, for in-services, and to talk to teachers from other schools. Her thinking and practice were also influenced by new ideas brought into her classroom by student teachers from the university who were placed in the school and supervised by [the PDS liaison]. In addition, Brandt reported that her principal and colleagues were supportive of her innovations. (p. 140)

> Brandt's team acted as a support system and sounding board and provided motivation for each other. Initially, the team discussed ways of grouping students and assigning teachers to subject matter. Over time, teachers on the team began to discuss individual students, curricular units, and themes . . Participatory management in which teachers were involved in making decisions helped her feel free to try out new ideas. (pp. 141, 142)

Many analysts of restructuring schools have noted that, as in other organizations, as teachers become increasingly involved in establishing a sense of direction and purpose for their school they have a greater tendency to feel responsible for their work. As Duke (1994) notes:

> The likelihood that teachers will embrace collective accountability for student learning is directly related to the extent to which teachers influence the formation of policies related to curriculum, instruction, evaluation, and other professional aspects of schooling. (p. 26)

Thus, engagement in decision-making supports teachers in taking professional initiative in the areas of curriculum and teaching and in accepting collective forms of professional accountability that reinforce shared standards of practice. All of these interrelated components of a professional role encourage teachers to take on more leadership and greater responsibility for improving teaching, schooling, and student learning.

Collective learning: Teachers as problem solvers and change agents

Because professional development schools explicitly aim to restructure teacher education and schooling simultaneously, teachers and teacher educators exert leadership as change agents for the profession as a whole as well as for their local schools. This process is often triggered as teachers inquire into their own schools and how they work for students. At Louisville's Fairdale High School, teachers' research coupled with a shared decision-making structure for problem-solving stimulated a major change process:

As part of a self-study, ten teachers followed ten children through a school day. When it was over, teachers said things like, "It was boring," or, "You know, this isn't a very humane place to be." Another teacher reported that no adult had spoken to the child she was following the entire day . . . Another activity that brought teachers together was reading about education and teaching. The teachers who went to Gheens Academy read and began to trade articles from *The Kappan, Educational Leadership*, and *Education Week* . . . Even before participative management was initiated at Fairdale, the teachers started changing things. In 1987 a steering committee consisting of elected teachers, students, administrators, support staff, and parents adopted operating procedures and set up task forces to study, design, and implement program changes generated from staff brainstorming sessions (Fairdale High School, 1990). The next year, Fairdale joined the Coalition of Essential Schools. Changes in pedagogy, use of time, and student testing have flowed from incorporating coalition principles into school practice . . . "Make no mistake about it," [the principal] said, "we are into culture building here. We are building a community culture outside and a professional culture inside." (Kerchner, 1993, p. 39)

Whitford (1994), who also studied Fairdale, adds that the principal has encouraged faculty invention and experimentation, including the formation of interdisciplinary teams, making scheduling changes, participating in hiring and budget decisions, and taking part in the University of Louisville's teacher preparation program. A key touchstone is that problem-solving starts from a focus on learners and their needs. The principal's guiding question in relation to these aspects of teacher leadership is, "Is it good for students?" One of Fairdale's teachers comments, "One of the best things about being here is getting to try things you've always wanted to try. We inspire each other to figure out how to work best with kids" (p. 80).

Shared decision-making in these PDSs supports an orientation toward schoolwide problem-solving and change. At Lassiter Middle School, another PDS associated with the University of Louisville:

One effect of participative management has been to create a situation in which teachers perceive how the entire school works. "It makes me question more than it used to," said one teacher. "We began to realize how connected things are . . . how setting up multiage teams affects the flow of students into other grade-level teams. It forces you to get involved in the workings of the school." (Kerchner, 1993, p. 36)

PDSs tend to bring a problem posing and problem solving frame to the practice of teaching as educators inquire together about what is working and what might work beter. At Wells Junior High, for example, the whole notion of staff development was turned on its head. The emphasis shifted from outside consultants to in-house experts. Collaborative learning groups replaced the traditional lecture/demonstration format. Problem posing and problem solving supplanted the recipes and prescriptions for effective schools that teachers had heard for years and never managed to implement . . . The staff embraced the emphasis on teacher decision-making and its connection to instruction. They resonated with the possibilities and connected their practical knowledge of children and classrooms with the professional research. There was a sense there was not a right answer. We began to see ourselves as problem solvers and we got a sense that if it doesn't work, we can always retool and change. (Miller & Silvernail, 1994, p. 31)

As faculties become empowered to pose and solve problems, they assume leadership for change from within, rather than looking upward or outward for leadership. The process is often a recursive one. As change is sought within classrooms and schools, some teachers take the initiative to look further, to look for opportunities to make needed changes at the district level and beyond. At Lark Creek Middle School:

Teachers believed that their colleagues have become "very aware of what's going on in education." One teacher commented that she had grown immensely through the opportunity to "look at education with different eyes," while others spoke positively of the constant flux of new ideas entering the building. A smaller group of teachers also spoke of their roles beyond the classroom and their desire to effect change in their district. (Grossman, 1994, p. 61)

Thus, teacher leadership emerges and grows rather than being appointed or assigned. It becomes the product of a fuller, more wide-ranging professional role. As suggested above and as described more fully below, a major dimension of that role is inquiry as a centerpiece of what it means to be a professional teacher. As centers of inquiry, PDSs develop this aspect of teacher leadership perhaps more fully than do other kinds of restructuring schools.

Teachers as researchers

Murray (1993) notes that a PDS is not a demonstration school, it is an inquiry school. This idea is echoed in the fact that, for example, the core beliefs of Lark Creek Middle School include the tenet that "central to the life of our PDC are dialogue and inquiry" (Grossman, 1994, p. 55). In a variety of ways, professional development schools encourage teachers to probe their practice both individually and collectively. Classroom-based as well as school-wide inquiry occurs as teachers participate in structured studies, pursue their curiosity where it leads them, help interns undertake case studies, and engage in action research. PDSs provide a number of different models for how schools can create the necessary conditions for teacher research.

Wells Junior High School is in a district that, with the support of the superintendent, takes seriously the notion that classroom teachers should lead school
change; that teachers should be committed to the practice of continual inquiry,
using the knowledge from research and theory to examine both their instruction
and practice and the structure within which they work in order to continually
improve the educational program for their students (Miller & Silvernail, 1994).
The PDS there started with staff development structured so that teachers could
examine research for its relevance to local problems. Over time teachers also
began to develop their own research:

> "Using the knowledge" became the starting point for developing a
> new view of staff development at Wells. The school staff redefined
> its use of the district's allocated workshop days. Rather than pro-
> viding time for formal presentations by outside consultants, the days
> were used for teachers' review of research and for critical discussion
> and reflection. For example, on one such day teachers spent 2 hours
> individually reading research about grouping. During another day,
> they worked in cooperative learning groups to share their percep-
> tions on the research they had read. On yet another day, the staff met
> to engage in the process of consensus building with the goal of reach-
> ing a decision about grouping practices in the school. (p. 30)

At Pontiac elementary in South Carolina, a teacher–researcher course was
designed by Pontiac elementary faculty members and a university professor. The
course is "designed to work with individuals and small groups and . . . to
address the needs teachers have regarding the investigation of curricular
approaches and their effects" (Berry & Catoe, 1994, p. 186).

At the Learning/Teaching Collaborative teachers have created a research
group that supports their inquiries into problems of classroom practice. As one
explains:

> I'm particularly interested in how kids collaborate in their writing,
> who brings what to the collaborative process, whether collaboration
> allows kids to take more or less risks with their writing . . And it's
> changed my philosophy because not only do we [the teacher research
> group] do classroom-based research, we also do a lot of reading in
> the research on linguistics. So I've read a lot of Vygotsky and
> Baktine. It's been incredibly interesting. (Boles & Troen, 1994, p. 16)

As another L/TC colleague noted, research provides another way for teachers
to look at kids, to understand how they function. In addition, some PDS teachers are working to extend their roles beyond the bounds of classroom inquiry
to enrich the broader field of research with teachers' perspectives. An L/TC
teacher explained:

> I would say my role as teacher researcher is definitely in the theater
> of leadership in the sense that I'm committed to developing a voice

for the teacher researcher in the context of the larger research world, in making that a viable voice that's different, yet heard in that context. (Boles & Troen, 1994, p. 17)

Lytle and Cochran-Smith (1994) argue that the teacher's voice can inform a broader reform agenda and can sensitize academic research while bringing it greater utility and legitimacy in the eyes of teachers:

> Limiting the official knowledge base for teaching to what university academics have chosen to study and write about contributes to a number of problems, including discontinuity between what is taught in universities and what is taught in classrooms, teachers' ambivalence about the claims of academic research, and a general lack of information about classroom life from the inside. It is widely agreed that instructional reform depends upon tapping into, and supporting, teachers' potential to be thoughtful and deliberate architects of teaching and learning in their own classrooms, and is contingent upon members of the teaching profession developing their own systematic and intentional ways to scrutinize and improve their practices. Teacher research may function as one of the critical ways that teachers access and interrogate their own knowledge and thus shape the larger reform agenda. (pp. 45–46).

Levine (1992) asserts that, in order for teachers to learn to cope with the uncertainty of their work, professional practice must be viewed not merely as craft-like, learned through apprenticeship, teaching by rule of thumb and imitation, but rather as involving reflection, experimentation, and inquiry. Lieberman and Miller (1992) add:

> In a school where teachers assume leadership in curriculum and instruction and where reflective action replaces routinized practice, providing opportunities and time for disciplined inquiry into teaching and learning becomes crucial. Unlike traditional school settings, professional practice schools are places where teachers, sometimes working with university scholars and sometimes working alone, do research on, by, and for themselves. Professional practice schools must provide the conditions that allow teachers to develop the skills, perspective, and confidence to do their own systematic investigation. (p. 108)

One can argue that, in fact, the professional teacher is one who learns from teaching rather than one who has learned how to teach, since the myriad puzzles posed by individually unique students always demand that teachers study their students and examine how different ones of them respond to different learning opportunities. Since "pre-service teacher education can only be an introduction to this career-long process of learning from teaching" (Snyder, 1994, p. 108), what it must impart to prospective teachers is the capacity to inquire sensitively and systematically into the nature of learning and the effects of their actions on learners. Understanding research about teaching enables teachers to learn from their own ongoing research on teaching. John Dewey (1929) put it this way:

> Command of scientific methods and systematized subject matter liberates individuals; it enables them to see new problems, devise new procedures, and in general, makes for diversification rather than for set uniformity. This knowledge and understanding render (the teacher's) practice more intelligent, more flexible, and better adapted to deal effectively with concrete phenomena of practice . . Seeing more relations he sees more possibilities, more opportunities. His ability to judge being enriched, he has a wider range of alternatives to select from in dealing with individual situations. (pp. 20–21)

Houston (1993) notes that if teachers investigate the effects of their teaching on students' learning they come to understand teaching "to be an inherently problematic endeavor, rather than a highly routinized activity" (p. 126). Engaging pre-service and in-service teachers in active inquiry in PDSs is one route to changing teaching practice and curriculum content on a profession-wide scale, by enabling teachers to engage in the kind of ongoing inquiry and problem-solving needed for learner-centered teaching and indepth understanding rather than mere coverage of subject matter.

The traditional view of teaching assumes a linear relationship between knowledge and practice, in which knowledge precedes practice, and the practitioner's role is limited to being either a user of research or the subject of it (Levine, 1992). Going beyond such a view, mature PDSs recognize the reciprocal, interactive nature of knowledge building. In these settings, knowledge is not viewed as a static entity residing in the upper echelons of the school bureaucracy, to be packaged in guidelines and directives and handed down from on high. And teachers are no longer seen as isolated transmitters of that knowledge to students. Instead, teachers' collaborative work with students, fellow teachers, pre-service teachers, and university faculty pushes them to explore and reimagine their own roles in the collective construction of knowledge about the learning and teaching process. As they learn and build knowledge jointly, they also transform practice as collective use of shared knowledge affects all participants, altering the shape of the school structure, and extending the knowledge base of the profession itself.

Challenges and possibilities for egalitarian teacher leadership

This vision challenges the hierarchical and positional conceptions of leadership current school structures embody. PDSs pose an alternative to the vertical system of advancement which identifies leaders as those holding formal occupational roles "above" those of teachers (McLaughlin & Yee, 1988). Boles and

Troen (1994) note that "teachers in PDSs are *expected* to exert influence beyond their classrooms and play important roles in the larger arena of the school, school district, and professional community" (p. 1). As teachers' expertise is recognized, their roles expanded, and their responsibilities increased, they not only become more powerful leaders, they become more powerful learners—and modelers of learning. As Lieberman et al. (1988) note:

> Teacher-leaders . . . are not only making learning possible for others but, in important ways, are learning a great deal themselves. Stepping out of the confines of the classroom forces these teacher-leaders to forge a new identity in the school, think differently about their colleagues, change their style of work in a school, and find new ways to organize staff participation . . It is an extremely complicated process, one that is intellectually challenging and exciting as well as stressful and problematic. (p. 164)

These examples suggest that, in contrast to bureaucratic forms of teacher leadership that simply create a few more slots in an already isolating and compartmentalized structure, PDSs—like other restructuring schools—can offer organic forms of professional leadership that develop intrinsically in connection with systemic organizational change within a school. In the course of restructuring, opportunities to collaborate and take initiative are available at every turn. The specific teacher leadership responsibilities that evolve are not prescribed a priori, but are varied, flexible, and idiosyncratic to individual school teams and their distinctive situations (Devaney, 1987, cited in Smylie & Denny, 1990). Embedded in this idea of teacher leadership is a core commitment to teacher decision-making and professional discretion and a belief that "the most desirable form of leadership, in fact, actually may be that which is not limited to particular roles, but instead derives from expertise and experience" (Duke, 1994, p. 25). As Boles and Troen report of the Learning/Teaching Collaborative:

> The PDS has established a new sub-culture in the schools that supports risk-taking, values leadership, and simultaneously maintains the norms of equality and inclusion among teachers. The PDS enables teachers to circumvent the more traditional school culture that does not reward, and often obstructs, risk-taking and collaboration. (p. 24)

Similarly, at I.S. 44 in New York City, the PDS has developed a climate within which leadership opportunities are widely available within a new definition of the professional role that focuses on professional responsibility for finding ways to succeed with students:

> It is a context within which teachers can make new meanings about children and schooling, and create new ways to foster important learning experiences for children. Rules are few, but decisions are many. The framing context is the professional model of learner-

centered teaching, of personal and collective initiative, responsibility, and accountability, a model at odds with the bureaucratic model of external rules, prescriptions, and evaluation (Darling-Hammond, 1990). (Lythcott & Schwartz, 1994, p. 127)

In these contexts, the greatest support for teacher leadership is not a formal title but restructured time and relationships that enable them to take on the leadership tasks they are ready to engage. A useful model may be that of the Learning/Teaching Collaborative, where participating teachers are allocated time instead of new titles. Just as teachers who work in more professionally-structured schools in many European and Asian countries use half or more of their time in enacting an expanded professional role—engaging in collegial planning and curriculum development, mentoring, and other tasks that are considered "leadership functions" in more bureaucratized settings (Darling-Hammond, 1990)—teachers in the L/TC are given a significant amount of Alternative Professional Teaching Time to be used at their discretion. New role relationships are made possible by the PDS collaborative governance structures, the commitments of L/TC schools to change, and the explicit mission of the partnerships with Wheelock and Simmons Colleges. The very different configuration of teaching time and responsibilities in other countries is a function of radically different organizational and personnel structures that invest in many more teachers and many fewer administrators and other adjunct personnel in schools (Darling-Hammond & Sclan, in press). In the long run, such transformations of time and teaching roles will require fundamental restructuring of school spending, staffing, and organization, rather than the dispensation of special privileges to a few.

These possibilities for teacher leadership have not emerged overnight and will not be expanded without struggle. The bounds and borders among the leadership forms described here vary from one professional development school to another, and each of them has had obstacles and setbacks along the way (Darling-Hammond, 1994b). Daly Lewis (1992) suggests that there is a developmental process through which school staffs work in transforming their schools into successful PDSs, and that this process involves moving along a continuum from rigid, top-down structures to more collegial, teacher-participatory approaches to staff development, decision-making, and leadership and vision for the school. In the course of this development, what may be most important is that they have access to images of what is possible and an openness to seize opportunities for participation, inquiry, and engagement in the continual rethinking of teaching and learning. As teachers grab hold of those emerging, unexpected opportunities, the possibilities for leadership may grow, and with them, the possibilities for a profession of teaching committed to ongoing inquiry and invention in support of student success.

References

American Association of Colleges for Teacher Education (AACTE) (1992). *Professional development schools: A directory of projects in the United States.* Washington, DC: Author.

Berry, B. & Catoe, S. (1994). Creating professional development schools: Policy and practice in South Carolina's PDS initiatives. In L. Darling-Hammond (ed.), *Professional development schools: Schools for developing a profession* (pp. 176–202). New York: Teachers College Press.

Boles, K. & Troen, V. (1994). *Teacher leadership in a professional development school.* Paper presented at the annual meeting of the American Educational Research Association, New Orleans, LA.

Daly Lewis, M. (1992). *Professional development schools: Conditions for readiness.* Paper presented at the annual meeting of the American Educational Research Association, San Francisco, CA.

Darling-Hammond, L. (1990). Teacher professionalism: Why and how. In A. Lieberman (ed.), *Schools as collaborative cultures: Creating the future now.* New York: Falmer Press.

Darling-Hammond, L. (1994a). Developing professional development schools: Early lessons, challenge, and promise. In L. Darling-Hammond (ed.), *Professional development schools: Schools for developing a profession* (pp. 1–27). New York: Teachers College Press.

Darling-Hammond, L. (ed.) (1994b). *Professional development schools: Schools for developing a profession.* New York: Teachers College Press.

Darling-Hammond, L. & Sclan, E. (in press). Who teaches and why: Dilemmas of building a profession for 21st century schools. In J. Sikula (ed.), *Handbook of research on teacher education.* New York: Macmillan.

Devaney, K. (1987). *The lead teacher: Ways to begin.* Paper prepared for the Task Force on Teaching as a Profession, Carnegie Forum on Education and the Economy. Berkeley, CA: Author.

Dewey, John (1929). *The Sources of a science of education.* New York: Horace Liveright.

Duffy, G.G. (1994). Professional development schools and the disempowerment of teachers and professors. *Phi Delta Kappan* 75 (8): 596–600.

Duke, D.L. (1994). *Drift, detachment, and the need for teacher leadership.* Paper presented at the annual meeting of the American Educational Research Association, New Orleans, LA.

Grossman, P.L. (1994). In pursuit of a dual agenda: Creating a middle level professional development school. In L. Darling-Hammond (ed.), *Professional development schools: Schools for developing a profession* (pp. 50–73). New York: Teachers College Press.

Houston, H.M. (1992). Institutional standard-setting in professional practice schools: Initial considerations. In M. Levine (ed.), *Professional practice schools: Linking teacher education and school reform* (pp. 124–132). New York: Teachers College Press.

Howey, K.R. (1988). Why teacher leadership? *Journal of Teacher Education* 39 (1): 28–31.

Kerchner, C.T. (1993). Louisville: Professional development drives a decade of school reform. In C.T. Kerchner & J.E. Koppich (eds.), *A union of professionals: Labor relations and educational reform* (pp. 25–42). New York: Teachers College Press.

Lemlech, J.K., Hertzog-Foliart, H. & Hackl, A. (1994). The Los Angeles professional practice school: A study of mutual impact. In L. Darling-Hammond (ed.), *Professional development schools: Schools for developing a profession* (pp. 156–175). New York: Teachers College Press.

Levine, M. (ed.) (1992). *Professional practice schools: Linking teacher education and school reform.* New York: Teachers College Press.

Lieberman, A. & Miller, L. (1992). Teacher development in professional practice schools. In M. Levine (ed.), *Professional practice schools: Linking teacher education and school reform* (pp. 105–123). New York: Teachers College Press.

Lieberman, A., Saxl, E.R. & Miles, M.B. (1988). Teacher leadership: Ideology and practice. In A. Lieberman (ed.), *Building a professional culture in schools* (pp. 148–166). New York: Teachers College Press.

Little, J.W. (1988). Assessing the prospects for teacher leadership. In A. Lieberman (ed.), *Building a professional culture in schools* (pp. 78–106). New York: Teachers College Press.

Lytle, S.L. & Cochran-Smith, M. (1994). Inquiry, knowledge, and practice. In S. Hollingsworth & H. Sockett (eds.), *Teacher research and educational reform* (pp. 22–51). Ninety-third Yearbook of the National Society for the Study of Education. Chicago: University of Illinois Press.

Lythcott, J. & Schwartz, F. (1994). Professional development in action: An idea with visiting rights. In L. Darling-Hammond (ed.), *Professional development schools: Schools for developing a profession* (pp. 126–155). New York: Teachers College Press.

McCarthey, S.J. & Peterson, P.L. (1993). Creating classroom practice within the context of a restructured professional development school. In D.K. Cohen, M.W. McLaughlin & J.E. Talbert (eds.), *Teaching for understanding: Challenges for policy and practice* (pp. 130–163). San Francisco: Jossey-Bass.

McLaughlin, M.W. & Yee, S.M. (1988). School as a place to have a career. In A. Lieberman (ed.), *Building a professional culture in schools* (pp. 23–44). New York: Teachers College Press.

Miller, L. & Silvernail, D.L. (1994). Wells junior high school: Evolution of a professional development school. In L. Darling-Hammond (ed.), *Professional development schools: Schools for developing a profession* (pp. 28–49). New York: Teachers College Press.

Murray, F.B. (1993). "All or none" criteria for professional development schools. *Educational Policy* 7 (1): 61–73.

Sergiovanni, T. (1987). The theoretical basis of cultural leadership. In L.T. Sheive & M.B. Schoenheit (eds.), *Leadership: Examining the elusive* (pp. 116–130). Alexandria, VA: Association for Supervision and Curriculum Development.

Smylie, M.A. (1994). Redesigning teachers' work: Connections to the classroom. In L. Darling-Hammond (ed.), *Review of Research in Education*, 20 (pp. 129–177). Washington, DC: American Educational Research Association.

Smylie, M.A. & Denny, J.W. (1990). Teacher leadership: Tensions and ambiguities in organizational perspective. *Educational Administration Quarterly* 26 (3): 235–259.

Snyder, J. (1994). Perils and potentials: A tale of two professional development schools. In L. Darling-Hammond (ed.), *Professional development schools: Schools for developing a profession* (pp. 98–125). New York: Teachers College Press.

Teitel, L. (1992). The impact of professional development school partnerships on the preparation of teachers. *Teaching Education* 4 (2): 79–85.

Tyack, D.B. (1974). *The one best system: A history of American urban education.* Cambridge, MA: Harvard University Press.

Whitford, B.L. (1994). Permission, persistence, and resistance: Linking high school restructuring with teacher education reform. In L. Darling-Hammond (ed.), *Professional development schools: Schools for developing a profession* (pp. 74–97). New York: Teachers College Press.

8

Learning about Organizational Learning

Coral Mitchell and Larry Sackney

Writing about organizational learning is a bit like heading into the Amazon rain-forest. You have some maps, and maybe a guide or two. You know the trip has the potential for exciting discoveries—some of which may lead to new and beneficial cures for some stubborn ills. But you also know that the terrain has been seen from many different perspectives and has been described in many different ways. And you know that the path is fraught with unanticipated circumstances and unexpected challenges. As Garvin (1993) laments, "the topic [of organizational learning] in large part remains murky, confused, and difficult to penetrate" (p. 78).

Part of the confusion may be due to the diverse and complex literature base. Dodgson (1993) attributes this diversity to fundamental disagreements over the meaning of organizational learning, but Huber (1991) takes a more optimistic approach. He argues that the disagreements may be due to an as yet inadequate development of the concept, and that, as future studies unfold, points of convergence will emerge. And some points of convergence are already apparent. In most cases, definitions refer to the ways in which members of an organization learn, individually and collectively, as they respond to demands for better organizational activity. At its best, organizational learning is a conscious and reflective approach to practice. Staff members, alone and together, assess the outcomes of their work, determine the kind of activity needed to meet the demands of the situation, and develop a learning experience that is most likely to yield the desired results. In essence, organizational learning leads to a change in how people behave toward their work, their organization, and one another.

Organizational learning has been discussed in the business and management literature for years, but only recently has it found its way into the educational literature. Although educators tend to speak in terms of "a community of learners" (Barth, 1990; Louis & Kruse, 1995), organizational learning is still the construct at the bottom of the discussions. But it is a construct that, for the most part, has not been extensively tested in the educational world. As Hargreaves (1995) points out, "If organizational learning is to help us in school renewal, we need to renew the concept in ways more suited to public school realities" (p. 18). In this chapter, we document our attempt to answer Hargreaves' challenge with a collaborative action research project that engaged a school staff in a program of organizational learning.

Into the school

Our concern at the beginning of this study was the utility of organizational learning for school personnel. Although we had found in the literature some advocacy for applying organizational learning in schools (e.g. Fullan, 1993; Louis, 1994), we found no examples of cases where school staffs worked explicitly with organizational learning behaviors. Consequently, we saw a need for empirical evidence of the effects of organizational learning practices in schools. We believed that the value of the concept could be effectively tested through an action-research design for at least three reasons. First, organizational learning is intimately concerned with improving current conditions, and we wanted to engage teachers and administrators actively in any attempt to improve the conditions that affected their work. Second, we were interested in re-shaping the notion of organizational learning to reflect the realities of life in school, and we believed that the most appropriate re-shaping would be done by practitioners working with the concept in their own context. Third, action research itself has been defined as an exercise in collective learning, whereby individuals test their own experiments, understandings, and insights in relation to those of their colleagues (Kemmis & McTaggart, 1988, pp. 8–9). In short, the two goals of action research, to improve practice and to generate knowledge, seemed ideally suited for testing the utility of organizational learning in an educational setting.

A prior question to be addressed was the place of organizational learning in the study. Much of the research before 1990 had used organizational learning as an object of study as researchers and scholars attempted to observe, analyze, and define the construct. However, in recent years researchers have used organizational learning constructs as a lens through which to examine other organizational issues or experiences (e.g. Leibenstein & Maital, 1994). Since much of the existing empirical work on organizational learning has been conducted in the business arena, we saw a need to define the construct for the educational world. Consequently, we chose to use organizational learning as an object of explicit study, rather than as a lens through which to interpret or to understand the results of research into some other issue.

Although action research is intended to fulfill the two-fold purpose of improving practice and of generating knowledge (Kemmis & McTaggart, 1988), researchers need to strike a balance between those purposes in order to avoid inherent tensions in their study (Hannay, 1989). The collaborative nature of this study suggested a solution through a division of ownership between the external researcher and the school staff. For the most part, the external researcher assumed ownership for the knowledge-generation tasks of providing theoretical information and of synthesizing the knowledge derived from the research experiences, and the internal participants assumed ownership for experimenting with learning processes and for improving practices.

Although the literature provides a comprehensive review of what organizational learning entails, little is said about how a group of people would go about "doing" it. Consequently, we generated some operational definitions in order to communicate with school personnel and to facilitate assessment of the nature of organizational learning in a school. A synthesis of the literature (Argyris, 1983; Argyris & Schön, 1978; Daft & Weick, 1984; Etheredge & Short, 1983; Friedlander, 1983; Fry & Pasmore, 1983; Mitroff, 1983; Schön, 1983; Senge, 1990) yielded the set of indicators of organizational learning listed in Table 8.1. We did not believe the behaviors listed in this table to be a comprehensive list, and we anticipated that further studies, including our own, would lead us to a different picture of the indicators. However, the indicators in the list provided a starting point for talking with teachers about organizational learning.

Armed with our list of indicators, and some preliminary definitions of organizational learning, we ventured into an elementary school to enlist a group of teachers in a discovery of what organizational learning might mean for them and of how the process of organizational learning would unfold in their school.

Table 8.1 Indicators of organizational learning—a literature-based synthesis.

1. Raising to awareness tacit assumptions and beliefs through reflective self-analysis
2. Engaging willingly in professional learning and growth
3. Understanding systemic influences and relationships
4. Sharing information openly and honestly
5. Developing a spirit of trust, empathy, and mutual valuing
6. Examining current practices critically
7. Experimenting with new practices
8. Raising sensitive issues and information
9. Understanding the inevitability of disagreement and conflict
10. Managing differences of opinion through inquiry and problem-solving
11. Engaging in dialogue in order to understand others' frames of reference
12. Changing frames of reference as warranted by team dialogue
13. Developing common understandings and language patterns
14. Developing a shared vision
15. Engaging in collaborative operation, planning, and decision-making practices
16. Correcting disruptive power imbalances

Shekina elementary school[1]

Shekina Elementary School, the site for the study, was situated on the edges of the inner core of a Saskatchewan university city. The school served a mixed community, with a number of families from the working class and a number of families on social assistance. High numbers of students came from single parent or blended families, and the population of the community was quite transient. Over the years, the incidence of troubled students had increased as many students with behavioral and learning difficulties had moved into the area. The frequent episodes of disruptive behaviors and the high student turnover conspired to make teaching at the school a challenging enterprise.

Twenty-one people staffed the school, including 16 teachers, one full-time administrator, and four support staff. All but two of the 17 professional staff participated in the research project. The teachers represented a wide spectrum of teaching experience and personality types. For the most part, school leadership was collaborative and teachers were involved in numerous school-wide activities. In the year prior to the study, the staff had initiated a school-wide behavior management program that aimed to reduce the disruptions to teaching and learning and to create a safer place for students and staff. Since they were continuing to learn about this program, it was used as the initial focus for understanding and applying organizational learning processes. However, as the study progressed, we applied the concepts and processes to emergent issues and difficulties in various aspects of school life.

Data collection and analysis

One of the first questions with which we grappled was the kind of research activities we would use to fulfill the purposes of the study. The heart of action research is a series of plan–act–observe–reflect cycles (Kemmis & McTaggart, 1988), but how we wanted those cycles to unfold was unclear at the beginning. In the early days, we spent considerable time discussing the sorts of activities that would be the most appropriate for this study and for this group of people. The external researchers were interested in activities that promoted individual and group reflection and professional conversation among the teachers, since both reflection and professional conversation have been identified as hallmark processes of organizational learning (Senge, 1993) and of action research (Kemmis & McTaggart, 1988). The school staff was interested in activities through which they could learn the most about organizational learning with the least disruption to their regular duties. Our deliberations led to the selection of six different research activities that we would pursue over a period of six months: theoretical information, individual interviews, large-group reflection meetings, interaction observations, verbatim transcripts, and data summaries.

To begin, the teachers needed some theoretical information about what was meant by the concept of organizational learning. Initial discussions revolved around the information functions operating within the school, and the nature

---

1 In order to protect the anonymity of the participants, the name of the school is a pseudonym.

of individual, group, and organizational influences on the teachers' collective processes. Later, the indicators of organizational learning listed in Table 8.1 were provided to the teachers, and these indicators became the "how-to" of organizational learning for the staff. From time to time, the teachers used them to assess the nature of collective processes in the school and to plan future processes and actions in harmony with the indicators.

Individual interviews were conducted with each participant at least once in order to gain insights into personal thoughts and experiences. Participants were invited to discuss their general experiences in the school, to reflect on their participation in the research project, and to explore the nature and the meaning of organizational learning as it was unfolding in the school. The interviews gave teachers opportunities to engage in self-reflection and to clarify their personal positions relative to school issues and needs. In addition to the survey interviews, iterative interviews were held from time to time with "key informants" (LeCompte & Preissle, 1993, p. 166) in order to share information, to check perceptions, to evaluate hunches, and to reflect on the direction and the value of the research project.

Large-group reflection meetings were held once a month to discuss the issues and needs arising from the previous month, to reflect on the nature of individual and collective learning that had transpired, and to plan directions for the next month. In the early months, the conversations focused on the individual and collective learning experiences of the staff and on the conditions that had both supported and inhibited their learning in the previous month. In later meetings, the group reflections centered around some current issue or need in the school. These reflection meetings, the embodiment of the plan–act–observe–reflect cycle, became the heart of the action research project for this group of teachers, for through these meetings the teachers began to develop some common understandings relative to school issues and to organizational learning.

Interactions among teachers during committee meetings and staff meetings were tracked and distributed to participants following the meetings. Committee meetings were commonly held at least once a week, and staff meetings once a month. We chose to track interactions primarily in relation to the ways in which the teachers spoke to one another and to the ways in which they dealt with school needs and problems during the meetings. In particular, we took note of the people contributing to group deliberations and the nature of those contributions. These observations served to raise awareness of communication patterns and power distribution among the staff.

Verbatim transcripts of individual interviews, committee meetings, and reflection meetings were distributed to the staff members who had participated in each activity. The teachers were encouraged to use the transcripts as fodder for their personal and group reflections. Participants used the transcripts to analyze their personal contributions and to reflect on the information and the interaction patterns evident in the transcripts. Furthermore, analysis of the conversations during committee and reflection meetings provided evidence of a shift in group processes relative to organizational learning as time went by.

Twice during the study, the staff was provided with a synthesis of the information derived from the reflection meetings and committee meetings. The first set of data was categorized according to the teachers' learning preferences and the conditions that affected their learning. The second summary was categorized according to the teachers' understanding of collaborative leadership and their thoughts on effective conflict resolution. These data summaries were instrumental in helping teachers move into the plan phase of the action research cycle. In other words, the teachers used the ideas in the summaries to guide their plans for future school activities, in relation both to professional development and to regular operations.

These activities, then, served as the mechanisms for the collection of qualitative data during the study. Data analysis was performed both by the teachers and by the external researcher. Alone and together, we conducted a deductive analysis using the indicators of organizational learning. The teachers analyzed the verbatim transcripts and the interaction checklists in order to assess the degree to which their learning processes reflected the behaviors in the list of indicators. In the early phases of the research project, analysis by the teachers was not common, but, as the months wore on, they became considerably more adept at analyzing their own practices in the light of the indicators. In addition to the deductive analysis, the external researcher conducted an inductive analysis of the data by using the constant comparison method (LeCompte & Preissle, 1993) of comparing data units with one another in order to find themes and patterns and by using conversation analysis in order to determine the nature of the discussions occurring during reflection, staff, and committee meetings. As data were compared, contrasted, ordered, and aggregated, organizing themes emerged.

The results

Data analysis confirmed the importance of the processes of reflection and professional conversation that had been embedded into the research design, and indicated the presence of two equally-important affective processes. Furthermore, conversation analysis revealed that the expression of these processes unfolded in three distinct but interrelated phases. As the teachers applied the principles and practices of organizational learning to a number of areas of school life, some assumptions developed about appropriate standards of conduct if organizational learning was to thrive. The story of the processes, phases, and standards of conduct of organizational learning, as identified at Shekina School, is presented in this section.

Processes of organizational learning

Although the indicators of organizational learning continued to serve as a definition during the entire study, the teachers gradually began to define the concept in terms that more closely reflected their own experience. Two teachers linked the concept to empowerment, effectiveness, and productivity, and another teacher defined it in terms of the development of sustainable organizational relationships. Other teachers talked about the development of a "positive work environment" and "a spirit of trust and understanding." Fundamental to many of the conversations about what organizational learning meant to these teachers was the notion of sustainable group processes in pursuit of greater effectiveness with colleagues and with students.

Data analysis revealed that the processes of importance were the cognitive processes of reflection and professional conversation and the affective processes of invitation and affirmation. The cognitive processes had been explicitly designed into the research activities, and they were crucial to the teachers' individual and collective learning. However, the affective processes emerged from the interaction norms that were established early in the project. These processes were influential in creating a climate in which reflection and professional conversation could flourish. Although we had worked at establishing the affective processes, we were surprised at the extent to which they influenced the cognitive processes. Since the four processes constituted the actual "doing" of organizational learning, some detail is in order.

Reflection emerged as teachers were encouraged, both in private interviews and in group meetings, to analyze critically their own practices, habits, beliefs, assumptions, and circumstances. In the early days of the project, critical reflection was difficult for the teachers, and examples of "aha" experiences seldom occurred. As the staff became more comfortable with analyzing their own practices in the light of the indicators of organizational learning, the insights became more personal:

> I don't want to impose my view, although I truly believe it's the right one. I have difficulty doing it when I'm—because I need other people's—because I am not—boy, it's tough to admit things you don't want to see about yourself . . . I'm afraid that I will alienate myself and I can't handle being alienated. I can't. I'm a people person and I need that commune and that feeling of interaction.

> Right now my head's bombarded with the things I do to inhibit [others]. If I learn to shut up and just give time to speak, that's the most critical thing.

These thoughts were shared in private interviews. Reflections in the public meetings were more related to group functions and school operations, and they, too, became more analytic and insightful as time went on. This comment came from a reflection meeting in a later month of the project:

> I was hoping that through the discussion of cases that maybe people could start positively reinforcing some of the students that are behavior problems and maybe offset some of those trips to the opportunity room, but I don't know if we really got far enough on that.

Reflection did not come naturally to most of the teachers, but, with the indicators of organizational learning as guides and with some "why" questions as prompts, they gradually became more reflective and analytic about their practices.

Professional conversation refers to in-depth discussions among staff members about professional issues, concerns, and ideas. The reflection meetings provided opportunities for teachers to engage in such conversations with one another. These conversations helped the teachers to develop common understandings, to become aware of one another's points of view, to raise and resolve issues, to make decisions, to learn about new possibilities for teaching, and to develop a cohesive team. The success of the reflection meetings indicated that the teachers placed considerable value on the opportunity to engage in professional dialogue with their colleagues, and their comments during these meetings emphasized the role that the conversations played in their individual and collective learning.

The teachers engaged in the process of invitation when they made overt efforts to draw other staff members into discussions or into collaborative practices. Most teachers in the study believed that the opportunity to contribute to group processes helped them to grow professionally, and they appreciated the times when they had been invited to share their ideas with other staff members. While most teachers felt they had adequate opportunities to participate in collaborative endeavors, not all teachers were equally involved. Certainly, the more vocal staff members were more visible in school operations and in staff discussions. However, as the study progressed, a number of teachers began explicitly to invite the quieter people to contribute to discussions.

Affirmation entails valuing the contributions of colleagues. It does not necessarily imply agreement. We found that we could disagree radically on a host of issues, but, in our disagreement, we acknowledged the value of others' opinions, ideas, or contributions. Several individuals commented that the honor given to their ideas helped to affirm them as professionals, and they saw this professional affirmation as important to their learning and to their participation in school life:

> When you know your input or your decisions are going to matter, then it's worth spending all that extra time—knowing that it's going to be appreciated in the end.

> You need to know you can speak and you will be heard . . . You need to feel that you are respected. You need to feel that what you say is valued. Not agreed with. That doesn't matter. But valued as a different opinion.

In their assessment of the research project, most teachers agreed that a significant outcome of the study had been an increase in their professional affirmation of one another.

In this school, the affective processes of invitation and affirmation established an environment in which the cognitive processes of reflection and professional conversation could unfold. Although we had not deliberately embedded invita-

tion and affirmation into the research activities, the norms that had been established in the school and that had characterized the early reflection meetings carried the spirit of the affective processes. As the non-threatening atmosphere
facilitated the teachers' participation in the cognitive processes, reflection and conversation moved beyond description and storytelling into analysis and evaluation.

Phases of organizational learning
Although the research project began on the first day of school following the
summer vacation, a few weeks passed before the staff was able to consider the
issue of organizational learning. Conversation analysis revealed that, once
underway, the process moved through three distinct phases. In the first phase,
which we called "naming and framing," discussions were characterized by
description, storytelling, and suggestion; in the second phase, which we labeled
"analyzing and integrating," by analysis and evaluation of current practices; and
in the third phase, which we called "applying and experimenting," discussions
revolved around implementation plans.

Naming and framing
In the first few months of the project, we were occupied with the tasks of building trusting relationships, developing common understandings about organizational learning, becoming familiar with the processes of reflection and
professional conversation, and identifying current practices. The indicators of
organizational learning became "touchstones" for understanding what we were
trying to accomplish in the research project, and the reflection meetings became
opportunities to hear the thoughts and feelings of our colleagues. An interaction
pattern that dominated early reflection and committee meetings was the tendency to share information, to make suggestions, and to express opinions. Only
occasionally would teachers ask for clarification or opinions from other staff
members. When differences of opinion arose in these early meetings, the teachers usually argued from their own position, with minimal attempts to understand the frames of reference of the colleague with whom they disagreed.

The first reflection meeting served to establish processes and norms that characterized the remainder of the research project. As the teachers became familiar
with the processes of reflection and professional conversation, some discomfort
regarding the research activities was eased, as seen in the following comments:

> I didn't know what to expect from the reflection meeting, being hon-
> est with you.

> It's nice just to talk and throw out ideas.

> I want to say that I really enjoyed this meeting, and this might be
> something we consider to do after Coral leaves.

In private conversations, several teachers commented that this meeting represented their first experience with professional reflection and that, in spite of
some initial hesitancy, they now felt more comfortable sharing their thoughts
and ideas with their colleagues. Of particular importance in raising the comfort

level were the norms of inclusion and affirmation, as all teachers were invited to share their ideas in the reflection meetings, and all contributions were accepted without judgement.

Although the teachers were becoming more familiar with the processes of professional conversation and reflection, their understanding of organizational learning was rudimentary during the first months of school. From time to time they used the indicators to reflect on their own experiences, but little evidence could be found of actual changes to practice in this first phase. However, the emerging understandings and relationships provided a necessary foundation upon which to build the more complex aspects of organizational learning.

Analyzing and integrating
In the fourth month of the project, we conducted a "mid-course evaluation" to assess what had been accomplished and what future directions we wanted to take. We realized that the large bulk of our activities had centered on building awareness of organizational learning and of current practices, and on creating a spirit of trust and mutual respect among staff, but we had not tackled any of the tougher indicators. In subsequent reflection meetings, the teachers became more analytic and thoughtful about their practices:

> I've noticed that, although we do have a beginning and an end, we get off topic really easy. We zoom out into left field and we chew around there for about a page and a half, and then, wham, we're back on again. A lot of divergent thinking is going on.

> We don't bring enough to closure where we add the subjective evaluation in our own minds, whether we're by ourselves or with colleagues.

During this time, the first evidence of expressing and testing assumptions emerged. A disagreement had arisen over plans for a school activity. After several minutes of wrangling, the principal stopped the conversation to inform the teachers about his assumptions and to inquire if they shared those assumptions. When they told him they did not, he replied, "I guess we need to clear up just what we're assuming, don't we?" From that point on, discussing and testing assumptions became a hallmark indicator of organizational learning at the school.

In the second phase, the teachers spent considerable time discussing their own learning preferences and the conditions that either supported or inhibited their learning. They began to take a critical look at a number of personal characteristics, staff interaction patterns, school structures, and program characteristics that affected the way in which they did their work. By this time, the processes of reflection and professional conversation were becoming integrated into school life. In private and staff-room conversations, teachers reflected on their experiences during the meetings, and they offered agenda items for future meetings. One teacher commented on the changes she had noticed:

> Some of the things that came out after the reflection meetings, and
> the talking . . . people were continuing to think about it. On their
> own accord they'd say something to you or to somebody else and
> you'd sometimes hear somebody make a comment to somebody else:
> "Gee, that was an interesting point you made." And they'd been
> thinking about it. And that doesn't always happen.

The tasks of the second phase of organizational learning were to integrate professional reflection and conversation into the life of the school, to examine and evaluate current practices and conditions, to express and test assumptions, and to consider new possibilities. The teachers now felt a commitment to organizational learning because they were beginning to see some benefits in terms of enhanced professional relationships and greater awareness of school issues and needs. Although we had not yet made any inroads into actual changes in practice, the information generated during the discussions both in committee and reflection meetings helped to move the teachers into the application phase.

Applying and experimenting
In a discussion with the principal, we realized that organizational learning could be applied to at least three aspects of school life: technical areas such as learning about new initiatives, cultural areas such as conflict resolution and group interactions, and political areas such as decision-making and leadership. We decided to distribute to the staff a summary of the information that had been generated in various meetings in order to make it more accessible for guiding the teachers' future planning. The first time the data summaries were put to explicit use was in preparing for a professional development workshop to be conducted by the behavior management committee. The following comment demonstrates how the data summary informed their planning efforts:

> It was very helpful for us to plan with . . . It was really interesting to
> see that some of us learn in different ways. Normally when you pre-
> pare for an in-service, you worry about getting content across but
> not always are you so in tune with what the people need or the
> avenues which they need it to be got across with. That's what was so
> nice about all the compilation, because we were able to say, "If you
> want people to learn, then this is what we need to include. If you
> want to avoid disgust or frustration or those types of things, then
> this is what we have to avoid doing."

Although the teachers had moved into the application and experimentation phase, application progressed on a broken front. They had analyzed technical aspects well, and were ready to apply what they had learned in that area of school life. However, they had not yet considered issues relative to the political and cultural aspects, and application in those areas moved forward more slowly.

During this time a change in the interactions in committee meetings became apparent. In previous meetings, teachers had largely advocated their own positions. Now they showed concern for the positions and opinions of others, and they began to ask for clarification and ideas in order to understand the frames of reference of their colleagues. Discussions in committee and reflection meetings continued to be analytic and evaluative in nature, with the added dimension of looking for ways to use the information they were generating. For example, in one reflection meeting, the staff discussed the issues of collaborative leadership and conflict resolution, and used the information to deal with a staff problem that had arisen. Although they experienced some difficulty resolving that particular issue, they believed that they had laid a solid foundation for dealing with difficult concerns in the future. In the application and experimentation phase, then, the staff began to use the information they had generated in the first two phases in order to plan school activities and to deal with school issues. Organizational learning came to life for the teachers as they applied what they had learned to their pursuit of more effective practices.

The gestalt of the phases

For this group of teachers, organizational learning became the process by which they developed common understandings about a particular issue or aspect of school life, analyzed current practices in the light of those understandings, and experimented with new possibilities. Interestingly, different indicators of organizational learning gained importance in the different phases, as outlined in Table 8.2. As each new issue or need arose, the teachers moved through the three phases, but the process did not follow a linear pattern. Rather, it was a cyclical and dynamic process with each phase unfolding from and enhancing the other phases. Furthermore, they engaged in different phases at the same time,

Table 8.2   Indicators and phases of organizational learning.

Phase	Indicators
Naming and framing	Developing a spirit of trust Developing common understandings Developing a shared vision Sharing information openly and honestly Engaging in collaborative practices Engaging in professional learning and growth
Analyzing and integrating	Using reflective self-analysis to raise awareness of assumptions and beliefs Examining current practices critically Understanding the inevitability of conflict Engaging in dialogue in order to understand the frames of reference of others Raising sensitive issues for discussion
Applying and experimenting	Experimenting with new practices Changing personal frames of reference if warranted Managing differences of opinion through inquiry and problem-solving

depending on the need. For example, as the teachers moved into applying and experimenting in the technical area, they continued with naming and framing, and analyzing and integrating in the political and cultural areas.

Nor did the process flow unceasingly from start to finish. At times, the teachers were distracted by a busy event calendar or by stressful school circumstances. For some teachers, the reflection meetings provided an opportunity to gain support and encouragement from their colleagues during such times, and they were able to regain a sense of perspective and to "regroup" in order to meet the next challenges. For others, however, the busy or stressful times caused them to disengage from the research project and from organizational learning. Furthermore, because they were constructing their own understandings of organizational learning, only gradually did some clear patterns emerge. In the early months of the project, they equated organizational learning with the reflection meetings, and only used the principles during those times. Later, a number of teachers began to apply the indicators of organizational learning to other aspects of collective activity, but a few staff members never made the transfer into the larger life of the school. In short, the ebb and flow of organizational learning followed the ebb and flow of life in the school.

Standards of conduct

As the study progressed, various individuals began to discuss the standards of conduct that they thought were essential to the development of effective organizational learning. In essence, the staff believed that organizational learning was dependent upon all members maintaining psychological safety for their colleagues, taking personal responsibility for the success of the school, and honoring diversity among individuals. These standards of conduct were not attributed only to the leaders, but were seen to be important standards for all staff members to follow.

These teachers were clear about the importance of feeling safe from threat, retaliation, embarrassment, or punishment. They told stories of how an unsafe environment in previous situations had stifled their participation and their learning. They indicated that respectful treatment of colleagues increased the level of trust, and that all teachers deserved equal respect and consideration. This standard of conduct did not imply the hiding of sensitive issues, but, rather, the respectful treatment of colleagues while those issues were being resolved.

> Knowing colleagues' beliefs, emotions, feelings, help us to deal respectfully and kindly and effectively when resolving conflict or differences. Conflict is inevitable and should be thought of simply as an essential, necessary process towards growth and understanding.

The degree of trust among the teachers had a considerable impact on their individual and collective learning, because it affected the degree to which they were willing to work with and learn from their colleagues. These teachers believed that the trust could be developed only if they treated one another with care and respect.

The teachers were equally clear about the need for every staff member to take responsibility for supporting school tasks and activities. This support was defined in active terms. Teachers were expected not just to give verbal or moral support, but also to be actively participating in whatever needed to be done at the school level. They indicated that their collective efforts were enhanced when all staff members participated in school events, and that their efforts were hindered when some teachers failed to become involved in school life outside the classroom or failed to fulfill their staff responsibilities. From time to time stories were told of how the lack of responsibility on one person's part had increased the load for someone else, and had led to resentment, frustration, and alienation. This standard of conduct was one we struggled with for the duration of the project. Although the reflection meetings were instrumental in bringing some of the less involved teachers into the group, a small handful of teachers never joined the collaborative processes. Their absence was noticeable, and called our attention to the need for each individual to assume some level of responsibility for collective efforts.

The theme of diversity among teachers was discussed frequently during the study. Certainly, the concept that each individual is unique is not a new thought. However, the staff came to believe that differences should not just be recognized, but should be honored, valued, and encouraged. The acceptance of diverse opinions and styles emerged during the reflection meetings and proved to be helpful in furthering the teachers' individual and collective learning. For example, individuals acknowledged that the level of acceptance for differences had helped them to feel comfortable sharing their ideas during a workshop on positive reinforcement:

> It was so broad that I think people felt good when it got to teacher programs to explain exactly what they were doing. Some things that may have been more or less out in left field, here people could tie into it and say, "I do this in my classroom."

> You saw it was everybody's own style, and you didn't think, "Oh, gee, I'm not doing anything like that in my classroom. I'm not going to say what I'm doing." Everybody was able to say, and in the same way, everybody accepted all the different things they were doing.

While the teachers agreed on the importance of sharing a common goal, they also believed that diverse expressions of that goal would enhance school operations and would give them alternative perspectives from which to learn. As they progressed through the study, many teachers commented on how differently they now viewed the world, as a consequence of the diversity of ideas being presented in group deliberations. In other words, they discovered that an awareness and appreciation of diversity was a necessary element for organizational learning to occur.

Discussion

In our exploration, we found that attention to the processes, the phases, and the standards of conduct did not guarantee the existence of effective organizational learning. A review of the literature had suggested that collective learning was shaped by three constellations of influences: conditions related to individual members, to group interactions, and to organizational structures (e.g. Argyris & Schön, 1978; Hedberg, 1981). Our experiences confirmed the influence of these three sets of conditions, and added a fourth category: influences related to specific contextual conditions. These four sets of influences had a considerable impact on the degree to which different individuals participated in group processes and contributed to collective learning. We cannot provide a detailed explication of the influences here, but we mention them in order to point out the conditional nature of organizational learning. Kolb (1984) argues that learning progresses in "fits and starts," and in our study, the fits and starts were shaped by personal, interpersonal, organizational, and contextual conditions.

Table 8.3 provides an overview of our discoveries about the essence of organizational learning at Shekina Elementary School. The teachers attributed a number of individual and staff benefits to their participation in organizational learning. From time to time, individuals commented on feeling a greater sense

Table 8.3.   Organizational learning in an elementary school.

Purpose	To create sustainable organizational processes To increase school effectiveness and teacher productivity
Phases	Naming and framing Analyzing and integrating Applying and experimenting

Processes	Cognitive: Reflection
	              Professional conversation
	Affective: Invitation to participate
	              Affirmation of colleagues

Standards of conduct	All members are responsible for school success Diversity among individuals is honored and valued Psychological safety is maintained during group deliberations
Content	Technical aspects of teaching and learning Political aspects of leadership and decision-making Cultural aspects of relationships and differences
Influences	Individual Group Organizational Contextual

of inclusion in school operations, greater awareness of staff patterns and of colleagues' thoughts, and increased personal reflection and self-awareness. At the staff level, the teachers participated in an increased number of professional conversations in a variety of settings. These conversations became progressively less descriptive and more analytic in nature. Closer personal relationships among staff members grew from the conversations, and several individuals commented on the group cohesion that had developed. An increased valuing of diversity was evident as committee and reflection meetings included more inquiry into the thoughts and opinions of others.

The development of organizational learning served at least three purposes in the school. First, it highlighted the importance of the staff role of the teacher. Through their conversations and reflections, the teachers uncovered some assumptions guiding their own and their colleagues' behaviors. They began to examine their professional relationships with their colleagues, to consider the ways in which those relationships affected teaching and learning, and, consequently, to think about the responsibility they owed to their colleagues. Second, the teachers began to see connections between their own learning and the learning of their students. The group deliberations gave the teachers some new possibilities to consider, and the conditions supporting organizational learning helped them to experiment safely with those possibilities. Consequently, their teaching, and thus the learning of their students, was felt to be more effective. Third, the process of organizational learning provided a framework for school assessment and school improvement. The reflective conversations offered a forum for raising emergent issues or for suggesting future directions. Affirmation and invitation served to bring the thoughts of most teachers into the discussion, thereby raising awareness of diverse perspectives. The indicators of organizational learning helped the teachers to analyze and evaluate current practices in the school, to highlight some areas that warranted improvement, and to increase confidence in those aspects that worked well. Consequently, the teachers found the experiences to be instrumental in enhancing their professional growth.

We do not claim that this research project turned the school into a learning organization. As some teachers observed, we had only "scratched the surface" of most of the notions that we came to associate with organizational learning. Although we could articulate our understandings, we had changed practice only to a small extent. But in our "muddling through the process," we gained some insights into how and why teachers might go about the practice of collective learning. We offer our discoveries as another perspective from which to understand how organizational learning might unfold in a school. In that light, we make the following observations.

Organizational learning is not necessarily doing something more, but rather doing the same things differently
An organization can only be said to have changed when the people who work within it behave differently. Organizational behavior arises through establishing

rules about working together and through "playing the game" within those rules (Reynolds, 1995; Sackney et al., 1995; Swieringa & Wierdsma, 1992). As we struggled to construct some meanings about organizational learning, we found that we were reflecting on and talking about the assumptions and beliefs underlying the teachers' practices, ideas, problems, and decisions. These underlying constructs served as rules by which the staff members were expected to behave in the school. Our reflections and discussions were instrumental in helping the teachers to analyze the effects of the rules on current behaviors, and to develop, or at least to propose, more effective practices, decisions, solutions, and ideas. In essence, we were exploring ways to approach the day-to-day needs, concerns, and experiences of classroom teachers collectively rather than individually. In so doing, we were constructing more effective rules by which to live and work.

Organizational learning flourishes within a collaborative school culture
Closely associated with the rules of conduct is the notion of the school culture. Sackney and Dibski (1994) distinguish between individualistic and collaborative cultures. They maintain that individualistic cultures serve to isolate teachers from one another, to suppress important information, and to promote the status quo. By contrast, a collaborative culture is characterized by "an atmosphere where joint planning, collegiality, support, trust, experimentation, creativity, and reflection are the norm" (p. 4). Such characteristics are clearly more supportive of organizational learning than are those of an individualistic culture. In this study, a collaborative culture emerged when the teachers followed certain standards of conduct, that is, when each individual took responsibility for maintaining psychological safety in group deliberations, for participating in school-wide activities, and for honoring the diversity among them. Teachers expressed considerable frustration when others failed to live up to these norms. We do not suggest that these norms reflect universal standards to be followed in each situation. However, they do indicate that collective processes are deeply affected by the manner in which individual teachers conduct themselves in relation to the group. In that regard, they signal a need for teachers to talk about the standards of conduct that would support a collaborative culture in their school.

A related question of concern is the extent to which schools are characterized by collaborative cultures. In an extensive review of the literature on collaboration in schools, Macleod (1994) uncovered a continuum of cultural characteristics moving from isolation and individualism (most schools) through contrived collegiality and small-group collaboration (some schools) to a community of learners (very few schools). In her review, she found much advocacy for collaborative school cultures (e.g. Barth, 1990; Fullan & Hargreaves, 1991; Sergiovanni, 1994), but little concrete evidence of authentic collaboration that generated a true community of learners. Her own study of four middle-years schools supported that conclusion. In each school, she found a deeply entrenched and pervasive culture of individualism that promoted independent work, "parallel play," "trick-trading," and "materials sharing," but little meaningful collaboration. Although the literature presents a somewhat pessimistic view of school cultures in relation to collaborative practices, we believe that educators can and should begin discussions about the cultural norms in their school and about the standards of conduct that would promote movement toward a collaborative culture.

Organizational learning implies collaborative leadership of and authentic involvement in school operations

Not only cultural but also political aspects of school life (Corbett & Rossman, 1989) are implicated by the results of this study. Shekina School was characterized by power-sharing among the staff members. Most teachers felt free to raise issues that affected teaching and learning and to suggest ways to improve school operations, and they felt a sense of ownership for school directions and decisions. However, Fullan and Hargreaves (1991) point out that power-sharing, at any level, is not evident in most school systems. Rather, school systems are often characterized by authoritarian leadership and unilateral decision-making (Sackney & Dibski, 1994). The teachers at Shekina were convinced that the processes of reflection and conversation would be difficult to pursue with an autocratic leadership because of the fear of punishment. Furthermore, we became aware that power imbalances do not always involve the school administrators. In this study, the teachers who controlled the process also controlled the outcomes. In many meetings, certain individuals were seldom heard. Since their opinions and wishes were not voiced, they may not have been reflected in the outcomes. Of course, not all teachers wish to be involved in collective processes, but that choice should be freely made, not forced by the dominance of other colleagues.

A related issue has to do with the extent to which teachers' contributions make a difference in the final outcomes. These teachers distinguished between authentic involvement and tokenism. They knew their involvement in school decisions and directions was authentic because they saw their input reflected in the final outcomes. However, they were skeptical as to how much impact their contributions would have in larger system decisions, and several of the teachers had chosen not to participate in system committees. If teachers are invited into a collective activity but their personal contributions are not valued, it becomes a hollow process, and they are likely to disengage. Therefore, attempts to establish effective organizational learning in schools need to honor the knowledge and the contributions of individual teachers.

Organizational learning requires a culture that encourages diversity and fosters dialogue

One of the most unexpected results of the study was the importance of valuing the diversity among staff members. The existing literature on organizational learning has not focused on the concept of diversity, and we were unprepared for the powerful influence of this issue. At Shekina School, the valuing of diversity provided an environment in which the teachers felt safe to express their own point of view and to explore alternative positions that they had not previously considered. We believe that, if organizational learning is to flourish, leaders need to ensure that diverse expressions of common goals are not just tolerated, not just accepted, but valued and encouraged, and that differing views and practices become a cornerstone of group dialogue.

Organizational learning involves complex interactions among an array of cognitive and affective processes
Although reflection and conversation had been explicitly designed into the research activities, the affective processes emerged from the group norms that had been established during reflection meetings, and the importance they assumed in supporting the cognitive processes came as a surprise to us. Reflection and conversation have previously been linked to organizational learning (e.g. Senge, 1993), but affirmation and invitation have not. We found that reflections and professional conversations flourished when the teachers affirmed one another as professionals and when all staff members were invited into the deliberations. The affective processes provided a safe climate for discussing the "undiscussable issues" (Argyris, 1983), and the cognitive processes offered a means for resolving the issues. In our view, attempts to generate effective organizational learning need to pay attention to both the cognitive and the affective domains.

We also found a close connection between reflection and conversation. Schön's (1983) idea of the reflective practitioner has been widely discussed in the literature, but this notion has largely been described in terms of the private reflections of the individual. Some have argued that critical inquiry is most effective when reflection and conversation are combined (Day, 1993; Foster, 1986; Osterman, 1990), and our study supports that argument. In this school, the group discussions provided grist for the private reflection mill. Different people provided different pieces to each puzzle, and the teachers gained new insights, perspectives, and interpretations as they listened to the contributions of various staff members. Furthermore, the private reflections of individual teachers enhanced the quality of the group conversations. The teachers gradually moved through descriptions and stories to reach a level of analysis and evaluation in their conversations that had not been evident in the early days of the study. The dynamic interaction between reflection and conversation appeared to help these teachers move into "double-loop learning" (Argyris & Schön, 1978) whereby they began to wrestle with their own assumptions. Although we cannot claim to have discovered all the important processes involved in organizational learning, our results do highlight the complexity of the interactions among the processes.

Organizational learning is a dynamic enterprise that unfolds through a series of interdependent phases
The phases of organizational learning identified in this study provide an operational framework for engaging individuals in collective learning. In many situ-

ations, problems arise when decisions and judgements are made before adequate time has been invested in discussing issues and developing common understandings (Roemer, 1991). In other words, beginning the "applying and experimenting" phase without first "naming and framing" or "analyzing and integrating" may lead to decisions or solutions that are inappropriate or untimely. In this study, the naming and framing phase served to clarify positions and opinions. As the teachers moved beyond advocacy of personal positions toward inquiry into the positions of their colleagues, they were able to broaden their perspectives and to reach agreement. The knowledge gained during naming and framing provided a solid foundation upon which to move into less familiar and, perhaps, more sensitive modes of interacting. Furthermore, the analyzing and integrating phase opened up new ideas for possible experimentation. Each phase served critical functions in the overall process of organizational learning, and none should be overlooked in any attempt to engage members in new initiatives.

The phases of organizational learning raise questions regarding the ways in which decisions are made in schools. Whether decisions are made unilaterally or collaboratively, the importance of the naming and framing phase should not be overlooked. In this study, decisions, programs, and resolutions were most effective when the staff understood the reasons for them. Decisions or programs that seemed arbitrary were likely to be received unenthusiastically, at best, or with hostility, at worst. Teachers need opportunities to discuss options or to consider the reasons for a particular choice. We wish to emphasize that framing is different from receiving. If teachers are simply told something, understanding has not necessarily taken place. Rather, framing is an active process that includes discussion and reflection. Consequently, teachers need time to talk about issues and to reflect on the discussion in order to reach a comfort level with decisions or directions that affect their lives.

Opportunities for organizational learning need to be structured explicitly into the teachers world of work
Although the teachers in this study took time to reflect alone and together when the study was in progress, they did not continue with reflection meetings after the project ended. They understood the benefits of organizational learning, but their hectic classroom schedules took them away from one another. If organizational learning is indeed instrumental in improving the instructional repertoire and the knowledge base of teachers, then time to engage in conversation and group reflection needs to be built in to the teachers' work day. However, many teachers already feel overwhelmed, and an additional task is not likely to be welcomed. Louis (1994) argues that enabling teachers and school administrators to work together will require "a profound change in the use of time" (p. 17) in contemporary schools. For the most part, the classroom role of the teacher is the only one that receives attention in the educational world. Although the classroom functions are critical and should not be neglected, teachers' classroom duties can be enhanced when attention is paid to the staff role of the teacher and to the issues of staff learning and staff relationships. Time spent in collective

learning is not time wasted. Rather, it provides a solid foundation for improved instruction, and it deserves to be explicitly scheduled into the teachers' world of work, and certainly not to be tacked on at the end of the day.

Action research methods have the power to foster organizational learning
Action research is a form of self-reflective inquiry through which educators work to improve their own practices, their understanding of those practices, and the situation within which the practices are performed (Carr & Kemmis, 1983, p. 152). The action research spiral of plan–act–observe–reflect–plan is, at its heart, an organized process of learning that moves participants from action to retrospective understanding to prospective action. The close connection between reflection and dialogue indicates that this type of learning is most insightful and empowering when undertaken collectively (Day, 1993; Kemmis & McTaggart, 1988). Personal and collective reflection and dialogue can inform the discourse of the practitioners and can contribute to a collaborative reconstruction of current practices. Members collectively transform practice into praxis (Carr & Kemmis, 1983) as they honor the authentic knowledge of their colleagues, as they encourage distinctive points of view, and as they engage in critical deliberations about practice. In essence, action research strategies support organizational learning through the development of more inclusive meanings, more productive connections, more sustainable processes, and more effective practices.

Conclusion

We began this investigation in order to develop some frameworks by which teachers might successfully cope with the many internal and external demands of their work. Although we did not explicitly address the issue of educational change, our experiments with organizational learning brought to the forefront some fundamental problems with many change efforts. We found that the teachers were more committed to efforts that addressed a specific need in the school, but we realized that many initiatives are not school-driven nor are they sensitive to individual school contexts. We found that the teachers learned better when they had opportunities to construct their own understandings about that which they were learning, but we realized that external initiatives seldom give teachers sufficient opportunities to develop personal understandings. We found that the teachers were better able to participate in organizational learning when school events were under control, but we realized that their participation in collective learning held the potential to de-stabilize the school and to push them into unfamiliar and, perhaps, uncomfortable places. We found that learning flourished when someone took responsibility for explicitly scheduling learning opportunities, but we realized that the teachers had little time to squeeze another task into their schedule.

These issues are not new. The question is whether or not organizational learning can shed any light on them. In our meandering through the process, we came face to face with a number of assumptions that guided the teachers' behaviors in the school, and we realized that we did not all view the world in the same way. We glimpsed the transformative power of organizational learning when we began to question our own assumptions and to open our hearts and minds to the points of view of our colleagues. Organizational learning calls us to take a curious approach to the world, a caring approach to the students and teachers with whom we work, and a problem-solving approach to the challenges of contemporary schools. We do not claim to have found all the answers to organizational learning or to educational change. We do claim to have made a step on the journey toward understanding how and why teachers might engage in collective learning. We have not charted the entire rainforest, but we have opened a path through the undergrowth.

References

Argyris, C. (1983). Productive and counterproductive reasoning processes. In S. Srivastva (ed.), *The executive mind* (pp. 25–57). San Francisco: Jossey-Bass.

Argyris, C. & Schön, D.A. (1978). *Organizational learning: A theory of action perspective*. Reading, MA: Addison-Wesley.

Barth, R.S. (1990). *Improving schools from within*. San Francisco: Jossey-Bass.

Carr, W & Kemmis, S. (1983). *Becoming critical: Knowing through action research*. Victoria, Australia: Deakin University Press.

Corbett, H.D. & Rossman, G.B. (1989). Three paths to implementing change: A research note. *Curriculum Inquiry* 19 (2): 163–190.

Daft, R.L. & Weick, K.E. (1984). Toward a model of organizations as interpretation systems. *Academy of Management Review* 9 (2): 284–295.

Day, C. (1993). Reflection: A necessary but not sufficient condition for professional development. *British Educational Research Journal* 19 (1): 83–93.

Dodgson, M. (1993). Organizational learning: A review of some literatures. *Organization Studies* 14 (3): 375–394.

Etheredge, L.S. & Short, J. (1983). Thinking about government learning. *Journal of Management Studies* 20 (1): 41–58.

Foster, W. (1986). *Paradigms and promises: New approaches to educational administration*. Buffalo, NY: Prometheus.

Friedlander, F. (1983). Patterns of individual and organizational learning. In S. Srivastva (ed.), *The executive mind* (pp. 192–220). San Francisco: Jossey-Bass.

Fry, R.E. & Pasmore, W.A. (1983). Strengthening management education. In S. Srivastva (ed.), *The executive mind* (pp. 269–296). San Francisco: Jossey-Bass.

Fullan, M. (1993). *Change forces: Probing the depths of educational reform*. London: Falmer.

Fullan, M. & Hargreaves, A. (1991). *What is worth fighting for? Working together for your school*. Toronto: Ontario Teachers' Federation.

Garvin, D.A. (1993). Building a learning organization. *Harvard Business Review* 71 (4): 78–91.

Hannay, L. (1989). Just who owns action research? In T.R. Carson & D.J. Sumara (eds.), *Exploring collaborative action research: Proceedings of the Ninth Invitational Conference of the Canadian Association for Curriculum Studies* (pp. 105–117). Jasper, Alberta, Canada.

Hargreaves, A. (1995). Renewal in the age of paradox. *Educational Leadership* 52 (7): 14–19.

Hedberg, B. (1981). How organizations learn and unlearn. In P.C. Nystrom & W.H. Starbuck (eds.), *Handbook of organizational design* (vol. 1, pp. 3–27). New York: Oxford University Press.

Huber, G.P. (1991). Organizational learning: The contributing processes and the literatures. *Organization Science* 2 (1): 88–115.

Kemmis, S. & McTaggart, R. (1988). *The action research planner* (3rd ed.). Victoria, Australia: Deakin University Press.

Kolb, D.A. (1984). Experiential learning. Englewood Cliffs, NJ: Prentice Hall.

LeCompte, M.D. & Preissle, J. (1993). *Ethnography and qualitative design in educational research* (2nd ed.). San Diego: Academic Press.

Leibenstein, H. & Maital, S. (1994). The organizational foundations of X-inefficiency. *Journal of Economic Behavior and Organization* 23 (3): 251–268.

Louis, K.S. (1994). Beyond "managed change": Rethinking how schools improve. *School Effectiveness and School Improvement* 5 (1): 2–24.

Louis, K.S. & Kruse, S. (1995). *Professionalism and community in schools.* Newbury Park, CA: Corwin.

Macleod, C.J. (1994). *An investigation of the perceptions and experiences of professional collaboration among teachers in middle years settings.* Unpublished master's thesis, University of Saskatchewan, Saskatoon, SK.

Mitroff, I.I. (1983). *Stakeholders of the organizational mind: Toward a new view of organizational policy making.* San Francisco: Jossey-Bass.

Osterman, K.F. (1990). Reflective practice: A new agenda for education. *Education and Urban Society* 22 (2): 133–152.

Reynolds, C. (1995). In the right place at the right time: Rules of control and woman's place in Ontario schools, 1940–1980. *Canadian Journal of Education* 20 (2): 129–145.

Roemer, M.G. (1991). What we talk about when we talk about school reform. *Harvard Educational Review* 61 (4): 434–448.

Sackney, L.E. & Dibski, D.J. (1994). *Collaborative school cultures under school-based management: Reality or myth?* Paper presented at International Congress for School Effectiveness and Improvement, Melbourne, Australia.

Sackney, L., Walker, K. & Hajnal, V. (1995). *Organizational learning, leadership, and selected factors relating to the institutionalization of school improvement initiatives.* Paper presented at the annual meeting of the American Educational Research Association, San Francisco.

Schön, D.A. (1983). *The reflective practitioner: How professionals think in action.* New York: Basic Books.

Senge, P.M. (1990). *The fifth discipline: The art and practice of the learning organization.* New York: Doubleday.

Senge, P.M. (1993). *Cornerstones of the learning organization: Videoconference.* Oak Brook, IL: The AED Foundation.

Sergiovanni, T.J. (1994). *Building community in schools.* San Francisco: Jossey-Bass.

Swieringa, J. & Wierdsma, A. (1992). *Becoming a learning organization.* New York: Addison-Wesley.

Part III

Foundations of Organizational Learning in Schools

9

Team Learning Processes

Kenneth Leithwood

Among the many initiatives associated with school restructuring, undoubtedly the most pervasive is site-based management (Murphy & Beck, 1995). But to acknowledge that this is the case contributes only marginally to our understanding of what is actually transpiring in schools claiming to be "doing it." Variation in the extent to which SBM is implemented is one source of confusion about what "doing it" actually means. But an even more basic source of confusion can be found in the multiple, legitimate models of SBM available to be implemented. Murphy and Beck (1995) identify three such models: administrative control SBM, in which the principal retains primary decision-making power; community control SBM, in which parents and other community members dominate school-level decision-making groups, such as school councils; and, professional control SBM. In the latter of these three models, school staffs as a whole, and teachers in particular, play the central role in decision-making.

All forms of SBM assume increased use of some form of participatory decision-making. But community and professional control models are less tolerant, by design, of mere consultation by the principal with others in the school (a not uncommon practice), as compared with an administrative control model. Furthermore, community control models, with their reliance on school councils, are expressly designed to ensure that the rights of community stakeholders are honored in schools, a largely political goal for participatory decision-making. In contrast, the purpose for professional control SBM appears to be improved educational practice, with fewer, other competing reasons. In the context of such an SBM model, the chances seem theoretically better that the efforts of participants will actually foster the individual and collective learning needed to improve the quality of instruction in the school. This may explain the results of a recent meta-analysis of empirical evidence concerning SBM effects, which Leithwood and Menzies sum up as follows:

> Professional control SBM appears to have more positive effects on the practices of teachers than either of the other two forms, and no more negative effects . . . Unexpectedly, as well, review evidence suggests that professional control SBM is the most likely form of SBM to increase professional accountability to parents and the wider community. (in press)

The evidence reviewed by Leithwood and Menzies indicates that usually the positive effects of SBM are undetectable: like Mohrman and her colleagues (1995), they suggest that SBM improves school performance only in the company of other important organizational conditions. These are conditions which support and directly foster the sort of individual and collective staff growth encompassed by the term "organizational learning" (OL). Such learning, then, is a vital matter to understand if we are to better appreciate what is entailed in successful school restructuring. In particular, it is important to understand how collective learning occurs in the many task forces, groups, committees and teams that are responsible for enacting the bulk of non-classroom business in restructuring schools. Exploring the meaning of team learning (as an instance of organizational learning) and the conditions which foster or inhibit its development is the purpose of this chapter.

Much has been written about group work already, of course. (e.g., Brightman, 1988; Goodman, 1986; Hackman, 1991; Worchel et al., 1992). These sources, however, have a decidedly prescriptive and managerial cast to them. As McGrath (1986) notes, what is largely missing is useful theory for understanding team learning and for aiding in the interpretation of empirical evidence. In particular, McGrath argues that "we must study [and build theory to explain] work groups . . . as intact social systems, and do so at a group level of analysis" (p. 368).

This chapter begins to address these limits on our understanding of team learning. The next section of the chapter takes up two problems that challenge the credibility of organizational learning as a useful organizational concept: whether it is reasonable to speak of a collective "mind" (McGrath's "intact social system") and, if it is, what is the nature of the collective learning processes that nourish such a mind. Building on the conceptual tools developed in that section, the following section offers a relatively detailed framework for guiding much-needed research on team learning processes in schools.

Collective learning processes

Limits on cognitive accounts of collective learning

A critical theoretical question raised by the distinction between individual and collective learning is whether the same conceptual tools reasonably can be expected to explain each. Much of the OL literature assumes this to be the case. Gioia (1986), for example, writes of "organizational cognition" as a multi-level phenomenon, not to be construed as something that happens only at the level of the individual. As Gioia's assertion suggests, theories of individual human cognition often are used to understand the learning of multi-person units (e.g., Hedberg, 1981). Whether literal or metaphorical, this use seems problematic on two counts.

Collective learning is not the sum of individual learning. Simon (1996) has argued that all learning takes place inside individual heads. The learning of an organization, from this perspective, is possible only through the learning of its individual members or through the ingestion of new members who have knowledge not already possessed by existing members. We should not, Simon claims, reify the organization by talking as though it knew something ("anthropomorphizing" it, to use Kim's [1993] term). Rather, we should be concerned about where knowledge in the organization is to be found and who possesses it, a position that would seem to dismiss the concept of collective mind.

Simon's position is problematic on two grounds. First, it asserts only one of two possible conceptions of knowledge—personal or internalized knowledge. This, of course, ignores objectively stored, externalized knowledge, the type codified in a book, a computer database, or an organizational policy manual, for example. Such knowledge resides outside any particular head and may reside also (at least personally meaningful versions of it) inside some individual heads.

A second problem with Simon's position is its failure to recognize that while the location and retrieval of knowledge in an organization is clearly an important part of understanding OL, it is separate from the processes used in its acquisition (more about these processes below). This is a not often recognized distinction between organizational knowledge and organizational learning and it is the latter of the two that is of greatest interest in this paper.

There are several additional problems associated with conceptualizing collective learning wholly as the product of multiple individuals' learning. Attempting to conceptualize and explain collective learning primarily as multiple individual learning runs the logical risk of (to borrow Holyoak's term) "combinatorial explosion" (1995, p. 271). This refers to the exponential and completely unmanageable increase in the complexity of an explanation based on multiplying the complexities of individual learning by the number of individuals in the organizational unit.

Conceptualizing collective learning as multiple individual learning also seems to have little potential for explaining variation in group performances. Such variation appears eventually to entail adaptations of individual learning and behavior in response to novel problems solved first by the collective group. For example, based on his study of a ship navigation team's response to a power failure under critical circumstances, Hutchins (1991) claims that "the solution was clearly discovered by the organization itself before it was discovered by any of the participants" (p. 14). While it is true in this case that the solution (what Simon referred to as knowledge) can be described in terms of changes in the understandings and consequent behaviors of individual team members, how the solution was learned requires a description of the interaction that took place among the individuals.

Cognitive theory is sometimes applied to understanding collective learning not as a means of providing a literal explanation of collective learning but as a stimulating metaphor for adding to our understanding of collective learning. An obvious example of such use is Morgan's (1986) image of organizations as brains; a more subtle use is Cohen and Bacdayan's (1996) analysis of organizational routines as if they were individual procedural memories. And as these analyses illustrate, metaphorical uses can provide powerful insights about collective learning. Such evidence notwithstanding, it is important to be clear on the limitations of this approach. Cook and Yanow (1996) note that using theories of individual learning to explain collective learning:

> raises a set of complex arguments concerning the ontological status of organizations as cognitive entities—specifically, arguments about how organizations exist and how the nature of their existence entails an ability to learn that is identical to or akin to the human cognitive abilities associated with learning . . . [To do this] one must first show how, in their capacity to learn, organizations are like individuals. (p. 435)

It is not clear, they argue, why two things so different in other ways as individuals and organizations should be expected to carry out the same processes in order to learn. Additionally, cognitive conceptions of individual learning are very much under development and, in many respects, contested. While they have contributed a good deal to our appreciation of how learning occurs among individuals, they cannot be adopted uncritically, even as metaphors, for insights about OL.

> Linking our understanding of organizational learning to cognitive theory, at the very least, obligates us to account in organizational terms for developments in that theory or to explain why this is not necessary. (Cook & Yanow, 1996, p. 436)

The team as a unit of collective learning

OL is a multi-level phenomenon. It takes place in many different organizational "units" (or across a continuum of different "organizational minds"). At one end of this continuum is the individual who learns within the context of the organization; at the other end is the collective learning of the whole organization. In the midpoint of these extremes is the team, with variations on these discrete units ranging from dyads of organizational members to multiple teams.

The remainder of this chapter is largely concerned with team learning processes because teams, as has already been noted (Brightman, 1988; Eisenstat & Cohen, 1991), increasingly are used for accomplishing organizational work. Teams, it is claimed, are more apt to represent the range of interests in an organization than is an individual; they may produce more creative solutions than an individual; and, their members and associates of these members are more likely to understand and support decisions made through participation in such decisions. Furthermore, communication is likely to improve among members when they are meeting together regularly. Team participation also is believed to be a valuable developmental experience.

Widespread use of teams is as evident in schools as it is in other types of organizations. Especially as schools struggle to restructure themselves, cross-role teams increasingly are expected to do the work of the organization, in particular the work of change and adaptation (Louis & Miles, 1990). Sometimes the membership of, and role relationships within, these teams remain informal; sometimes, as in the case of site councils for example (Murphy & Beck, 1995), membership and role relationships are well specified. Under both sets of circumstances, considerable ambiguity remains around general purposes, specific goals and working procedures; and neither set of circumstances lend themselves to the idea of a team within a conception of the organization as a bureaucracy.

From a rational or bureaucratic perspective (Scott, 1987), a team is a group of three or more people pursuing a specific set of goals within the context of a formalized set of social structures. Restructuring schools are explicitly moving away from this perspective on organizations, however, toward a view of organizations as open systems, especially open to interaction with their parent and wider communities (Conley, 1993). An open system is "a coalition of shifting interest groups that develop goals by negotiation; the structure of the coalition, its activities, and its outcomes are strongly influenced by environmental factors" (Scott, quoted in Weick, 1995, p. 70).

This open system definition of an organization begins to capture much of what is important about the nature and functioning of teams in restructuring schools. More specifically, it provides some conceptual purchase on the nature of team learning when teams' "nets of collective actions" (Czarniawska-Joerges, quoted in Weick, 1995, p. 74) are viewed as teams' collective minds, as discussed below. One net of collective action is distinguishable from another by the kind of meanings and products socially attributed to a given team.

At least for research purposes, however, it is important to have a more concrete set of criteria for identifying team, since teams vary enormously in the extent to which their members actually learn together in an interdependent manner. Because the paper is concerned centrally with such interdependent learning, Hackman's (1991) three criteria for defining a team are useful:

- the team is real (it is an intact social system with clearly defined boundaries between members and non members, and members are dependent on one another for the achievement of some sort of shared purpose)

- the team has one or more tasks to perform (there is an outcome of the team's work which is clearly accessible and for which the team has some collective responsibility)
- team members work in an organizational context (this means that they must manage relations with other individuals or groups in the larger social system, such as the parent organization)

The mind of the team

For the concept of team learning to be viable, it is useful to have a concept of a team mind which is doing the learning. How could such a collective mind be conceptualized? Weick and Roberts (1996) provide one useful approach to this thorny theoretical problem. As they explain:

> Our focus is at once on individuals and the collective, since only individuals can contribute to a collective mind, but a collective mind is distinct from an individual mind because it inheres in the pattern of interrelated activities among people. (1996, p. 334)

This solution builds on efforts by Wegner and associates in which:

> [team] mind may take the form of cognitive interdependence focused around memory processes . . [P]eople in close relationships enact a single transactive memory system, complete with differentiated responsibility for remembering different portions of common experience. People know locations rather than the details of common events and rely on one another to contribute missing details that cue their own retrieval. (cited in Weick and Roberts; 1996, p. 332)

As this explanation makes clear, only individuals can contribute to a collective mind and only the mind of individuals can be conceptualized as a set of internalized processes controlled by a brain. Collective mind must be an external representation; in Weick and Roberts (1996) terms, mind as activity rather than mind as entity. The team's mind, then, is to be found in patterns of behavior that range from being "intelligent" to "stupid."

Within contemporary cognitive science, there are two quite different ways of conceptualizing individual mind and these are related quite differently to Weick and Roberts' (1996) concept of collective mind. The longest standing cognitive science, as well as folk view of individual mind, according to Bereiter and Scardamalia (in press), is "mind as container." According to this view, individual mind consists of a set of internal cognitive structures which represent the personal meanings individuals associate with external events. From this perspective, mental activity involves the manipulation of knowledge objects and learning is the process of getting these knowledge objects into the mind. Furthermore, the actions that result from internal knowledge representations and the processes through which they are adapted are a product of, rather than a part of, individual mind. Collective mind, in contrast to this view, is defined by patterns of collective action. These actions are a consequence of individual cognitive structures and meanings.

Some of these structures and meanings are shared by all or many members of the team. Some, however, are unique to individual members and constitute the interdependent contribution of the individual mind to the collective mind.

A second and more recent conception of individual mind is offered by connectionism. Mind, from this perspective, can be knowledgeable without containing knowledge:

> The new metaphor suggested by connectionism is mind as pattern recognizer . . . The idea is that the mind aquires abilities and dispositions to recognize and respond in various ways to various patterns, but the patterns are not in the mind. We can say that the patterns are in the environment or, more cautiously, that the patterns are a way for us as observers to describe relations between the mind and the environment. (Bereiter & Scardamalia, in press, 489–490)

This view of individual mind actually is very close to Weick and Roberts' (1996) concept of collective mind in the emphasis it places on patterns which are both external to the brain and not dependent on internal representations. As is evident in the next section, connectionist views of individual learning also parallel defensible accounts of collective learning.

Team learning as mutual adaptation

Each member of a team is challenged to learn two things. One of these is a more or less shared understanding of the team's purposes and the sorts of actions for accomplishing those purposes permitted by the larger organization. When this learning challenge extends beyond explicit goals and procedures to include core values, beliefs, and behavioral norms, it is best viewed, as Cook and Yanow (1996) have argued, from a cultural perspective. This perspective draws attention to three important aspects of team learning.

First, a cultural perspective draws attention to one important aspect of the content of teams' learning: this is a team or corporate culture (e.g., Banner & Gagné, 1995), the collective programming of the mind in Hofstede's (1991) terms. Such a culture strongly shapes the nature of the interactions that occur among members of a team. For example, Schoenfeld (1989) claims that the value of a "local epistemological stance" shared by members of his research team contributed significantly to the synergy required for productive team learning. Second, a "cultural" perspective on team learning draws attention to one particular type of knowledge which teams usually need to draw upon and to acquire. For the most part, this is tacit knowledge (Cook & Yanow, 1996; Polanyi, 1967) and, as a consequence, difficult to consciously manipulate. Finally, a cultural perspective on team learning raises questions about how best to communicate such knowledge to other members of the team, perhaps new members or members who have difficulty understanding the shared meanings held by their team colleagues. Brown and Duguid (1991) show how the use of organizational stories to communicate tacit cultural, as well as more operational, knowledge influences members' cognitions and attitudes: the concreteness of stories facilitates the formation of mental imagery and its inherent interest attracts more attention and so more thorough processing.

The second learning challenge faced by all team members is what actions to take as individuals in order to contribute to the collective learning of the team. Collective learning develops from the actions of individuals as those individuals begin to act in ways heedful of the "imagined requirements of joint action" (Weick & Roberts, 1996, p. 338). These requirements might be implied in the corporate culture. But they could also include more immediate and explicit demands on joint action such as the demands, documented by Hutchins (1991), arising for a navigation team from a total loss of power on their large ship as it entered port.

March (1991), Hutchins (1991), and Schoenfeld (1989) describe the type learning in which individual members of teams engage as "mutual adaptation," an essentially connectionist view of individual learning applied to the team unit. Mutual adaptation can be of two sorts. One sort is largely unreflective. For example, in the case of Hutchins' (1991) navigation team, when individual team members confronted a change in what they believed was required of the whole team, each of the members adapted their usual contribution to the team as best they could, hoping that other members would be able to do whatever else was required. This was implicit negotiation of the division of labor. When it seemed not to be sufficient as a response to the team's new challenge, individual members then attempted to recruit others to take on part of what was assumed to be her part of the team's job, a second form of mutual adaptation.

In the case of Schoenfeld's (1989) research team, the task was to construct a coherent explanation for a set of data about a student's mathematical learning processes. Individual members or subgroups of the whole team typically constructed their own explanations of the data set first. Then they shared these explanations and engaged in some form of interaction which often produced a shared explanation significantly different than any of the individual members' or subgroup's explanations. Schoenfeld (1989) describes the process as follows:

> Let the subgroups be A and B, the idea [original explanation] in its old form X1, and in its improved form X2. The schema [process] goes as follows:
>
> • A either ignores or rejects X1 (so X1 would remain as it is);
>
> • B considers X1 important (but is unlikely to produce X2);
>
> • In group interactions, B convinces A to seriously consider X1; A suggests the change from X1 to X2, which the group ratifies; hence the group produces the change from X1 to X2, while neither subgroup A or B would have done so by itself. (p. 74)

In both the Hutchins and Schoenfeld examples, an imagined new challenge for the team serves as the stimulus for individual team members to adapt their contributions to the team's actions. In this way, the individual is contributing to the learning of the team. As other team members adapt their contributions, not only in response to their sense of the team's new challenge but also in response to the responses of other members, each team member learns about the adequacy of her initial response and perhaps the need to adapt further. This is the way in which the individual learns from the team. And, as Schoenfeld (1989) explains, "the result of the group interactions extended significantly beyond the 'natural' sum of the contributions that could have been made individually by the people involved" (p. 76).

Conceptualizing team learning as processes of mutual adaptation appears to minimize the likelihood of significant learning, variously referred to as double-loop (Argyris & Schön, 1978), higher-level (Fiol & Lyles, 1985), and turn-around (Hedberg, 1981), for example. It also seems to favor, in Weick's (1995) terms, more conservative adaptation strategies involving the "exploitation" of well-established solutions as opposed to "exploration" strategies involving wider searches for solutions, greater risk-taking and more innovation. Yet, as Weick (1995) points out, organizations that rely too heavily on exploration "are likely to find themselves trapped in suboptimal stable equilibria" (1995, p. 102).

Conceptualizing team learning as mutual adaptation, however, does not turn a blind eye to the more significant forms of learning required to maintain the necessary balance between exploration and exploitation that can determine whether an organization survives or not. Multiple, mutual adaptations by team members, even of a quite conservative sort individually have combined effects on a team's learning capable of producing radical changes in its collective mind (patterns of action). Both the Hutchins (1991) and Schoenfeld (1989) cases demonstrate that this is possible. Lant and Mezias (1992) also demonstrated, in a series of computer simulations, that repeated cycles of small changes in orga-nizational routines, which they referred to as convergence, periodically result in major reorientations in those routines. Such "punctuated equilibria" are char-acterized by "simultaneous and discontinuous shifts in strategy, the distribution of power, the firm's core structure, and the nature and permissiveness of control systems" (p. 47).

An adequate conception of team learning, then, must be able to account for both the necessarily incremental nature of changes in individual capacities which require learning and the more rapid forms of adaptation periodically required of organizations if they are to survive. Mutual adaptation appears to meet both of these criteria.

A framework for inquiring about team learning processes

Building on the conceptions of collective mind and collective learning described in the chapter so far, this section describes some of the conditions influencing such learning. My starting point is the work by Neck and Manz (1994) aimed at explaining effective group functioning by extrapolating from a form of dys-functional group behavior which Janis (1982) labeled "groupthink." The framework indicates that team learning, conceptualized as mutual adaptation, results in patterns of action (or the team's collective mind). Such learning on the part of the team is directly influenced by the nature of the team's leadership. It is influenced directly as well by a set of conditions for learning that grow out of the team's collective culture, something team leadership may also influence.

Both team leadership and team culture are shaped, not only by team learning experiences but also by those more distal, in-school and out-of-school variables that are part of the larger framework described at the beginning of the paper for understanding the nature, causes, and consequences of OL.

According to this framework, the outcome of a team's learning is a pattern of action. This may be a change from an earlier pattern, or a decision to continue with an existing pattern after, for example, carefully weighing alternatives and finding that current patterns remain a sufficient response to whatever was the stimulus prompting the team's thinking. In agreement with some others then (Cousins, 1995), this framework does not define team learning exclusively in terms of changed behavior.

Defining the outcome of a team's work as a pattern of actions begs the question, however, of how to define a group's effectiveness, a critical question for most people engaged in team learning in schools. Hackman (1991) and his colleagues offer one solution that is consistent with how team outcomes have been conceptualized to this point. This solution defines group effectiveness along three dimensions: the degree to which the team's products (decisions, actions, and the like) meet the standards of quality, quantity, and timeliness of the team's "clients"; the degree to which the process of carrying out the work of the team enhances the capacity of the members to work together interdependently in the future; and the degree to which the group experience contributes to the growth and personal well-being of team members. These three dimensions are at least a proxy for, if not a direct reflection of, the effectiveness of a team's pattern of actions.

Patterns of action are the direct result of interrelationships among the individual cognitions of team members, characterized earlier as mutual adaptation. Based on their consideration of groupthink, Neck and Manz (1994) suggest that the productivity of these adaptive processes is most effective when the conditions for team learning ("teamthink") include: encouragement of divergent views; open expression of concerns and ideas; awareness of limitations and threats to the work of the team; recognition of members uniqueness; and discussion of collective doubts.

The extent to which the conditions which enhance team learning are manifest depends on the team's culture, defined by three sets of variables included in the Neck and Manz framework. One set is shared norms, beliefs, and assumptions. Neck and Manz include as important here dominant attitudes towards the team's work; the belief, for example, that problems are opportunities to overcome challenges rather than obstacles that will lead to failure. That these attitudes have an important effect on the team's thinking also is supported by evidence from studies of expert group problem-solving processes provided by Leithwood and Steinbach (1995).

A second set of variables included as part of team culture is team self-talk. For both individuals and groups, it has been suggested that self-talk can serve as a tool for self-influence directed at improving the personal effectiveness of members (Janis, 1982; Neck & Manz, 1994; Weick, 1979). Such talk could be aimed at putting social pressure on team members deviating from the group, as is the case in instances of groupthink. However, group self-talk also could focus on the importance of what Senge (1990) refers to as personal mastery—efforts by each team member to continuously improve the individual capacities they contribute to the collective effort. This seems likely to have quite positive effects on team learning.

Group vision is the final set of variables included as part of team culture. This vision provides a relatively coherent sense of the team's overall purpose as well as more immediate goals. When the vision is widely shared and understood, it ought to be a primary resource for the team in determining what it needs to learn. Evidence provided by Leithwood et al. (1995) suggests that more tacit and deeply imbedded assumptions about purpose and associated mission appear to be the main source of members' understanding of the team's vision.

The stimulus giving rise to the need for team learning is sometimes conceived of as a crisis or an otherwise fairly dramatic event such as a strike (e.g., Watkins & Marsick, 1993) which forcefully draws group members' attention to the need for new learning. Recent research in schools (see Chapter 4), however, suggests that quite a few things have the potential to act as a stimulus and that schools vary considerably in their sensitivity to these stimuli. Both individual and team learning in schools can be stimulated by relatively everyday events, such as ongoing attempts at incremental improvement. The mandate assigning a "task force" is another example of a common, if not routine, organizational event which also serves as a stimulus for team learning (Gersick & Davis-Sacks, 1991).

Evidence in support of the plausibility of the direct influence of leadership on the conditions for team learning is provided by some of the research my colleagues and I have completed on the nature of transformational ("expert") principals' and superintendents' group problem-solving processes (Leithwood et al., 1993; Leithwood & Steinbach, 1995). These studies describe school leaders facilitating the work of staff teams, for example, by ensuring that the knowledge of all members of the team is made explicit in team discussions, and by encouraging innovative and coordinated action on the part of team members. In addition, these leaders also work to surface all members' interpretations of the problem(s) to be solved by the team and to develop, with the team, as clear as possible an interpretation of the problem(s). These leaders have well-developed plans for collaborative problem-solving which they share with team members and they periodically synthesize, summarize, and clarify the progress of the team. These examples of expert team leadership seem likely to foster most of the conditions for team learning, although our studies did not explicitly test this claim.

Team leadership, however, often is provided by those other than school administrators. This argues for caution in assuming that a set of practices apparently acceptable to, and productive for, a team when exercised by an adminis-trator will be equally productive when used by others in possibly informal or temporary leadership positions. Formal leadership roles, like the principalship, typically are vested by others in the organization with expectations and power that are at least partly unrelated to the person in the role. This has a significant effect on the meaning that members associate with the practices of the formal leader; the same assertive behavior might be interpreted by team members as "decisive" when it is exercised by a principal or superintendent but "overbear-ing and presumptuous" when exercised by an informal leader.

Conclusion

The primary intent of this chapter was to develop a theoretical perspective on team learning processes, one that could aid in reflective practice and possibly guide subsequent research on these matters. Teams are becoming a pervasive locus of work in many different types of organizations, not least restructuring schools. The perspective on team learning which has been developed begins by locating the collective team mind in the patterns of action undertaken by the team as a whole. Collective team learning entails change in these patterns of action through processes of mutual adaptation. These are processes in which individual members adapt their contributions to the team partly in response to their understanding of the nature of the new challenge facing the team and part-ly in response to the responses of their fellow team members' actions. In this way the team's learning has the potential to both precede and to contribute to individual members' learning.

These properties of team learning closely resemble the meaning of a complex system:

> in which groups of agents seeking mutual accommodation and self consulting somehow manage to transcend themselves, acquiring col-lective properties such as life, thought, and purpose that they might never have possessed individually. (Woldorp, 1992, p. 11)

To the extent that a team behaves in this way, it is a nonlinear dynamical sys-tem, a very different conception of a system than, say, Peter Senge (1990) had in mind when he extolled the virtues of seeking coherence and consistency across different elements of a system. This is what linear systems do, eventual-ly reaching a point of stability and equilibrium. But non-linear systems seem a better conception of a human team, for surely complex "systems" which are individuals will interact in unpredictable ways if the basic processes giving rise to individual thought themselves behave in a non-linear fashion. As Woldorp explains:

> Our brains certainly aren't linear: even though the sound of an oboe and the sound of a string section may be independent when they

enter your ear, the emotional impact of both sounds together may be very much greater than either one alone (this is what keeps symphony orchestras in business). (Woldorp, 1992, p. 65)

So why would we expect the interaction of two or more brains to be other than non-linear unless a set of "rules" (norms, guidelines, etc.) were imposed on that interaction which forced linearity, such as the conditions associated with "groupthink" (Janis, 1982)?

Premised on these orientations to collective mind and collective learning, the specific framework described in this paper for guiding inquiry on team learning processes is a substantial adaptation of Neck and Manz (1994). Team learning is directly influenced by a set of conditions inherent in the social interactions of the group (e.g., encouragement of divergent views). These conditions are themselves shaped directly by leadership (e.g., expertise in managing group processes) and team culture (e.g., shared values), and indirectly by conditions provided by the larger organization (e.g., its structure) and the wider environment in which it is located.

References

Argyris, C. & Schön, D. (1978). *Organizational learning: A theory of action perspective.* London: Addison-Wesley.

Banner, D.K. & Gagné, T.E. (1995). *Designing effective organizations: Traditional and transformational views.* Thousand Oaks, CA: Sage Publications.

Bereiter, C. & Scardamalia, M. (in press). Rethinking learning. In D.R. Olson & N. Torrence (eds.), *Handbook of education and human development: New models of learning, teaching, and schooling.* Cambridge, MA: Basil Blackwell.

Brightman, H.J. (1988). *Group problem solving: An improved managerial approach.* Georgia State University: College of Business Administration.

Brown, J.S. & Duguid, P. (1991). Organizational learning and communities-of-practice: Toward a unified view of working, learning, and innovation. *Organization Science* 2 (1): 40–57.

Cohen, M.D. & Bacdayan, P. (1996). Organizational routines are stored as procedural memory: Evidence from a laboratory study. In M.D. Cohen & L.G. Sproull (eds.), *Organizational learning* (pp. 403–429). Thousand Oaks, CA: Sage Publications.

Conley, D.T. (1993). *Roadmap to restructuring.* University of Oregon: ERIC Clearinghouse on Educational Management.

Cook, S.D.N. & Yanow, D. (1996). Culture and organizational learning. In M.D. Cohen & L.G. Sproull (eds.), *Organizational learning* (pp. 430–459). Thousand Oaks, CA: Sage Publications.

Cousins, J.B. (1995). *Understanding organizational learning for educational leadership and school reform.* University of Ottawa, unpublished paper.

Eisenstat, R.A. & Cohen, S.G. (1991). Summary: Top management groups. In J.R. Hackman (ed.), *Groups that work (and those that don't)* (pp. 78–86). San Francisco: Jossey-Bass.

Fiol, C.M. & Lyles, M.A. (1985). Organizational learning. *Academy of Management Review* 10 (4): 803–813.

Gersick, C.J.G. & Davis-Sacks, M.L. (1991). Summary: Task forces. In J.R. Hackman (ed.), *Groups that work (and those that don't)* (pp. 146–153). San Francisco: Jossey-Bass.

Gioia, D.A. (1986). Conclusion: The state of the art in organizational social cognition. In H.P. Sims, D.A. Gioia and associates (eds.), *The thinking organization* (pp. 336–356). San Francisco: Jossey-Bass.

Goodman, P.S., and associates (1986). *Designing effective work groups*. San Francisco: Jossey-Bass.

Hackman, J.R. (1991). Introduction: Work teams in organizations: An orienting framework. In J.R. Hackman (ed.), *Groups that work (and those that don't)* (pp. 1–14). San Francisco: Jossey-Bass.

Hallinger, P. & Heck, R.H. (1996). Reassessing the principal's role in school effectiveness: A review of empirical research, 1980–1995. *Educational Administration Quarterly* 32 (1): 5–44.

Hedberg, B. (1981). How organizations learn and unlearn. In P. Nystrom & W. Starbuck (eds.), *Handbook of organizational design, volume 1*. New York: Oxford University Press.

Hofstede, G. (1991). *Cultures and organizations: Software of the mind*. Berkshire, UK: McGraw-Hill Europe.

Holyoak, K.J. (1995). Problem solving. In E.E. Smith & D.N. Osherson (eds.), *An invitation to cognitive science, volume 3: Thinking* (second edition) (pp. 267–296). Cambridge, MA: MIT Press.

Hutchins, E. (1991). Organizing work by adaptation. *Organization Science* 2 (1): 14–39.

Janis, I.L. (1982). *Groupthink* (second edition). Boston: Houghton Mifflin.

Kim, D.H. (1993). The link between individual and organizational learning. *Sloan Management Review* 35 (1): 37–50.

Lant, T.K. & Mezias, S.J. (1992). An organizational learning model of convergence and reorientation. *Organization Science* 3 (1): 47–71.

Leithwood, K. (1994). Leadership for school restructuring. *Educational Administration Quarterly* 30 (4): 498–518.

Leithwood, K. & Aitken, R. (1995). *Making schools smarter*. Thousand Oaks, CA: Corwin.

Leithwood, K., Jantzi, D. & Steinbach, R. (1995). An organizational learning perspective on school responses to central policy initiatives. *School Organization* 15 (3): 229–252.

Leithwood, K. & Menzies, T. (in press). Forms and effects of school-based management: A review. *Educational Policy*.

Leithwood, K., Menzies, T., Jantzi, D. & Leithwood, J. (1996). School restructuring, transformational leadership and the amelioration of teacher burnout. *Anxiety, Stress and Coping* 9: 199–215.

Leithwood, K. & Steinbach, R. (1995). *Expert problem solving*. Albany, NY: SUNY Press.

Leithwood, K., Steinbach, R. & Raun, T. (1993). Superintendents' group problem-solving processes. *Educational Administration Quarterly* 29 (3): 364–391.

Louis, K.S. & Miles, M.B. (1990). *Improving the urban high school: What works and why*. New York: Teachers College Press.

March, J.G. (1991). Exploration and exploitation in organizational learning. *Organization Science* 2 (1): 71–87.

McGrath, J.E. (1986). Studying groups at work: Ten critical needs for theory and practice. In P.S. Goodman and associates, *Designing effective work groups* (pp. 362–391). San Francisco: Jossey-Bass.

Morgan, G. (1986). *Images of organization*. Newbury Park, CA: Sage Publications.

Murphy, J. & Beck, L.G. (1995). *School-based management as school reform*. Thousand Oaks, CA: Corwin Press.

Neck, C.P. & Manz, C.C. (1994). From groupthink to teamthink: Toward the creation of constructive thought patterns in self-managing work teams. *Human Relations* 47 (8): 929–952.

Newell, A., Rosenbloom, P.S. & Laird, J.E. (1990). Symbolic architectures for cognition. In M.I. Posner (ed.), *Foundations of cognitive science* (pp. 93–131). Cambridge, MA: MIT Press.

Polanyi (1967). *The tacit dimension*. Garden City, NY: Doubleday.

Schoenfeld, A.H. (1989). Ideas in the air: Speculations on small group learning, environmental and cultural influences on cognition, and epistemology. *International Journal of Educational Research* 13 (1): 71–88.

Scott, W.R. (1987). *Organizations: Rational, natural and open systems* (2nd edition). Englewood Cliffs, NJ: Prentice-Hall.

Senge, P.M. (1990. *The fifth discipline*. New York: Doubleday.

Simon, H.A. (1996). Bounded rationality and organizational learning. In M.D. Cohen & L.G. Sproull (eds.), *Organizational learning* (pp. 175–187). Thousand Oaks, CA: Sage Publications.

Watkins, K.E. & Marsick, V.J. (1993). *Sculpting the learning organization*. San Francisco: Jossey-Bass.

Weick, K.E. (1979). *The social psychology of organizing* (2nd edition). Reading, MA: Addison-Wesley.

Weick, K.E. (1995). *Sensemaking in organizations*. Thousand Oaks, CA: Sage Publications.

Weick, K.E. & Roberts, K.H. (1996). Collective mind in organizations: Heedful interrelating on flight decks. In M.D. Cohen & L.G. Sproull (eds.), *Organizational learning* (pp. 330–358). Thousand Oaks, CA: Sage Publications.

Woldorp, M.M. (1992). *Complexity: The emerging presence at the edge of order and chaos*. New York: Simon & Schuster.

Worchel, S., Wood, W. & Simpson, J.A. (eds.) (1992). *Group process and productivity*. Thousand Oaks, CA: Sage Publications.

10

Intellectual Roots of Organizational Learning[1]

J. Bradley Cousins

Introduction

It has only been within the past decade or so that principles from the domain of inquiry known as organizational learning have been systematically applied to the problem of educational reform. While some authors have considered directly educational reform issues from an organizational learning perspective (e.g. Fullan, 1993; Louis, 1994), others have used such conceptual frameworks to guide their study of associated phenomena. Studies of knowledge use (Cousins & Leithwood, 1993; Shujaa & Richards, 1989), policy implementation (Leithwood & Dart, 1996; Wills & Peterson, 1992; Woods, 1993), collaborative applied research (Cousins & Earl, 1992, 1995, see also chapter 7; Whyte, 1991), leadership (Leithwood, 1992; Taylor, 1986), schools' response to environmental change (Levin, 1993; Levin & Ezeife, 1994), and school system performance indicators (Leithwood & Aitken, 1995) are among the more salient examples. While organizational learning has captured the attention of those interested in educational reform, a considerable body of knowledge dating back more than 30 years has developed outside of this field. The purpose of this chapter is to explore these intellectual roots of organizational learning applications in education and to consider their implications for learning schools.

1 This chapter is adapted from Cousins (1996). The research was supported by the Social Sciences and Humanities Research Council of Canada (No. 410–92–0983). The opinions expressed within do not necessarily reflect those of the Council.

Several significant and comprehensive reviews of the literature[2] and the appearance of two "special issue collections" in management studies academic refereed journals[3] bear testimony to the appeal and currency of this theoretical perspective in fields of inquiry outside of education. In this chapter I consider and synthesize knowledge about organizational learning produced largely outside of education and I summarize this knowledge in terms of the key concepts associated with organizational learning processes and effects and the organizational conditions and factors supporting the development of learning systems in organizations. I conclude by considering the implications of this knowledge base for educational leadership and reform.

This chapter is adapted from an extensive review (Cousins, 1996) published recently in the *International Handbook of Educational Leadership and Administration* . In that survey article, I reviewed 46 theoretical or conceptual papers and 52 studies reporting original data. While many of these articles originated in the United States, I intentionally searched international (databases) and focused mostly on work published since 1980. Among the conceptual papers that were located are six survey articles (see footnote 2) and several others that sought to enhance our understanding of organizational learning by proposing theoretical developments and supporting these with logical argument.[4] Of the empirical studies, over half published data collected in the public sector, for example: emergency service, health care provision, education, and rural development fields. Slightly fewer of the studies collected data within the private sector from managers and employees of consulting firms, manufacturing operations, and a variety of other industries including oil, technology-based, and insurance companies. A small number of the studies focused on "virtual corporations" or joint ventures put together for relatively short-term commitments. Five of the studies used simulated data generated by econometric models and one looked at a joint venture between a university and the business community.[5]

This chapter uses the information available in these sources, first, to clarify organizational learning processes and their effects, and, second, to consider implications for fostering organizational learning in schools.

Organizational learning processes and effects

Organizational learning capacity is a dynamic and complex phenomenon perhaps best understood by considering learning processes and effects as influenc-

---

2 Daft & Weick, 1984; Fiol & Lyles, 1985; Hedberg, 1981; Huber, 1991; Levitt & March, 1988; Srivastva, 1983
3 Journal of Management Studies, 1983, 20 (1) and Organization Science, 1991, 2 (1).
4 For example, Daft & Huber, 1987; Dery, 1986; Garvin, 1993; Kiesler & Sproull, 1982; Levinthal & March, 1981; Lovell & Turner, 1988; Lundberg, 1989; Saffold, 1988; and Tushman & Romanelli, 1985.
5 For further details about the sample of studies reviewed, see Cousins (1996).

ing each other in a reciprocal way. Ten distinct yet overlapping concepts appear to be deserving of attention. The first four of these are concerned with basic questions of how information and knowledge are encoded and represented within the organization. These are:

- principles of social learning
- organizational knowledge representation
- behavioral versus cognitive learning distinctions
- levels of learning within the organization

The next three dimensions are concerned with how knowledge is framed, stored, retrieved, and communicated within the organization. They are:

- system structural versus interpretive learning systems
- organizational memory
- knowledge management strategies

Two concepts are strategic dimensions associated with the ongoing development and evolution of an organization's knowledge base. These I refer to as:

- "experiencing"
- knowledge acquisition processes

A final dimension concerns the organizational routines and responses that limit organizational learning capacity, often referred to as "dysfunctional learning habits."

Social learning

Organizational learning is not merely the sum of learning of members of an organization. In fact, team learning is often considered to be the lowest common denominator, and individual learning, far from guaranteeing organizational learning, may actually prevent it. Simon (1991) reminds us that all learning necessarily takes place in human heads and, within organizations, learning occurs in the heads of existing individual members or new members ingested from outside. [*see Chapter 10 for another perspective—Ed.*]

> Human learning in the context of an organization is very much influenced by the organization, has consequences for the organization and produces phenomena at the organizational level that go beyond anything we could infer simply by observing learning processes in isolated individuals. It is those consequences and those phenomena that we are trying to understand here . . . human rationality is very approximate in the face of the complexities of everyday organizational life. (pp. 125–126)

Organizational learning, then, is social learning. Bandura (1977, 1986) depicts personal factors interacting with environmental events and behaviors as determinants of social learning. He argues that knowledge is represented in symbolic form as shared meaning rather than as details of discrete events. From this point of view, collective problem-solving is not reducible to individual problem-solving: decision makers are inclined to apply rules derived from past experiences or to model decision rules used by others. Huber (1991) notes that learning occurs if any organization members acquire knowledge that can change the range of potential behaviours of the organization. Greater learning occurs as more members obtain this knowledge, as more varied interpretations are developed, and as more members develop a uniform comprehension of various interpretations. While it seems likely that organizational responses to external influences, for example, are more likely to be realistic when informed by multiple perspectives from within the organization, de Geus (1988) reminds us that it is "not the reality that matters but the team's model of reality, which will change as members' understanding of their world improves" (p. 78).

Organizational knowledge representation

How is knowledge represented at the organizational level? Consistent with the foregoing discussion of social learning, shared interpretations of events and collective representations of knowledge are constructed through social interaction. Argyris and Schön (1978) distinguish between "theories-of-action" (espoused theories about operational relationships and their consequences within the organization) and "theories-in-use" (the implicit principles and assumptions governing actions). The latter may or may not be aligned with the former. Individual images and shared maps or representations of theories-in-use are formed within organizations and, according to the authors, learning occurs when a mismatch between what is expected and what actually happens, "error detection", occurs. This mismatch between expectation and observation has been referred to as a "triggering event" by others (e.g. Virany et al., 1992).

While organizational knowledge is represented internally as shared interpretations and "mental maps," some researchers have directed their interests toward the external representation of shared knowledge. Organizational knowledge is said to be embodied in organizational routines, policies and procedures. In Attewell's (1992) terms "the organization learns only insofar as individual insights and skills become embodied inorganizational routines, practices, and beliefs that outlast the presence of the originating individual" (p. 6). Leithwood and Aitken (1995) suggest that routines (roles, procedures, forms, conventions) are based on the interpretation of past events. Similarly Nystrom and Starbuck (1984) argue that cognitive structures "manifest" themselves in perceptual frameworks, expectations, world views, plans, goals, sagas, stories, myths, rituals, symbols, jokes and jargon" (p. 55).

It seems pivotal to differentiate between knowledge acquired through interaction with the organization's environment and knowledge created from within, since each will have implications for organizational structures and processes designed to enhance learning outcomes. March (1991) frames this as a distinction between "exploitation" and "exploration." While the former is reflected in refinement, choice, production, efficiency, selection, implementation, and execution, the latter is evident in search, variation, risk-taking, experimentation, play, flexibility, discovery, and innovation. Of course, both modes of learning, "adaptive" and "generative" (Senge, 1990a), compete for precious resources within the organization; March suggests that organizations learn from experience how to divide among the two. He proposes that exploration processes usually have less certain outcomes, longer time horizons, and more diffuse effects, and therefore adaptive processes usually improve exploitation more rapidly than exploration activities. However, March (1991) cautions that:

> since long run [organizational] intelligence depends on sustaining a reasonable level of exploration, these tendencies to increase exploitation and reduce exploration make adaptive processes potentially self-destructive. (p. 73)

Two other considerations are important concerning trial and error adaptive learning processes. First, some scholars suggest that trial and error learning is only sufficient provided that enough time is available and that enough information is available when needed. Time lag between trial implementation and error detection is a second important consideration. Sterman (1989) observed that the underestimation of time lag was prevalent among participants in a marketing simulation. Both he and Senge (1990b) underscore the importance of paying attention to both outcome and action feedback.

Behaviourial versus cognitive distinctions

Whether change in behavior is a necessary indicator of organizational learning also has been a matter for debate among theorists. According to Fiol and Lyles (1985), through interaction with the environment, cognitive systems and memories are developed; learning is a process of improving organizational actions through better understanding. But some contemporary theorists are less concerned with observable organizational actions as a necessary criterion for learning. For example, according to Huber (1991), "an entity learns if, through its processing of information, its *range of potential behaviors* is changed" (p. 89, my emphasis). This perspective is congruent with Daft and Weick's (1984) notion of cognition development as distinct from behavior development, the latter implying new responses and actions based on interpretations. Fiol and Lyles (1985) extend Daft and Weick's original conception by developing a contingency chart with cognitive and behavioral continua as the x and y axes, respectively. This 2×2 framework enables one to differentiate four distinct organizational outcomes, only two of which are considered by the authors to be instances of organizational learning:

- low on both dimensions (no change)
- high learning and limited influence on action
- high action without much thought
- high on both dimensions (optimal)

Contingencies 2 and 4 are considered the only legitimate instances of organizational learning. But debate concerning category 2 appears to be unresolved. On the one hand, support is derived from Huber's (1991) position that increasing the range of potential organizational behaviors is both necessary and sufficient as the minimum condition for learning. On the other hand, the conception of organizational learning developed by Argyris and Schön (1978), specifically the notion of the detection and correction of error, is somewhat at odds with this perspective. According to Argyris (1993) learning as the detection and correction of error is intimately connected with action for several reasons: a gap exists between stored knowledge and knowledge required to act effectively; contexts are constantly changing; and it is necessary to codify effective action so that it can be reliably repeated. The essence of organizational learning then becomes the production of valid "actionable knowledge" which can be implemented, tested rigorously, and refined.

Levels of learning

A fourth important dimension of learning processes and effects, one that is basic to discussions about organizational restructuring or reorientation, is now both widely accepted and time-honored. Argyris and Schön (1978) originally differentiated *single-loop* and *double-loop* learning, although others have applied different labels.[6] Single-loop learning, according to Argyris and Schön, is reflected in changes that preserve the central features of theory-in-use. It is relatively shallow and relies quite heavily on established organizational routines (Fiol & Lyles, 1985).

Double-loop learning, on the other hand, reflects much deeper penetration into underlying assumptions and beliefs within the organization. Simon (1991, pp. 131–132) puts the distinction nicely: "I distinguish sharply between learning that brings new knowledge to bear within an existing culture and knowledge that changes the culture itself in fundamental ways." Learning at this level implies surfacing and articulating shared mental models and enhancing "sense-making" through close and open examination of core values, assumptions, and beliefs (Argyris, 1993; Argyris & Schön, 1978; Lundberg, 1989).

Much of the empirical research appears to support an "either–or" conception of these two levels of learning, but Lant and Mezias (1992) provide data to the contrary. Their study suggests that "the same processes that lead to first-order learning and convergence can provide raw material for second-order learning andre orientation" (p. 64). That is to say, normal organizational routines sometimes provide equivocal experiences which lead to second-order learning and change.

Another aspect of levels of learning is what Argyris and Schön (1978) call *deutero learning*, the extent to which an organization can stand back and contemplate its own learning behaviors, or engage "meta-learning" processes.

---

6 For example, low level vs. high level (Fiol & Lyles, 1985; Frey, 1990); first order learning vs. second order learning (Lant & Mezias, 1992); organizational change vs. organizational development or transformation (Lundberg, 1989); adaptive vs. generative learning (Senge, 1990b).

Although conceptually elusive and difficult to study, deutero learning appears to capture the essence of an organization's learning capacity.

Storage, retrieval, and communication of knowledge

System-structural versus interpretive perspectives
Limits on human rationality in organizational decision-making have brought into question assumptions about the nature of organizational learning. Lovell and Turner (1988), for example, conceive of an organizational learning continuum ranging, on the one hand, from a "rational analytic paradigm" implying sensitivity to information and research, and, on the other, "disjoint incrementalism" marked by variable coherence and relative inexplicability of organizational phenomena.

Daft and Huber (1987) also identify two basic organizational learning systems (or perspectives on learning). One of these, the system-structural perspective, represents the organization as a system for transmitting data, thereby raising the importance of physical characteristics of messages such as the amount and frequency of information. This perspective assumes that organizations exist in an objective environment, that understanding leads to action, and that the acquisition and rational analysis of data is the route to achieving that understanding. It also assumes that users know how to use the acquired information. The alternative, according to Daft and Huber, is the interpretive perspective which assumes that the organizational system gives meaning to data. This perspective raises to prominence the equivocality of data and the environment and the importance of context in sensemaking. Learning is a consequence of discussion, shared interpretation, changing assumptions and trial-and-error activities. Both accounts of learning capture important processes and they are not mutually exclusive.

Several empirical studies have employed the interpretive framework in attempting to illuminate organizational learning processes. Two themes emerge from these studies. First, social processing and reflective dialogue appear to enhance the interpretive skills of organization members, quite apart from the development of their knowledge. Second, organizational learning capacity may be enhanced through ongoing problem identification, interpretation, and responding.

Organizational memory
Central to any conception of learning, effects, and processes (individual or organization) is the concept of memory. Levitt and March (1988) define organizational memory as the rules, procedures, technologies, beliefs and cultures that are conserved through systems of socialization and control. Cohen (1991) proposes that organizations have repertoires of activities for acquiring information and improving what they can do. He suggests that "building and modifying the repertoire are fundamental activities because they embody learning in routines,

thus constituting a major form of organizational memory" (p. 135). Organizational memory is intimately related to the process of encoding information—specifically, determining the "newsworthiness" of incoming information (Tiler & Gibbons, 1991). Organizational memory is represented by the structures within which knowledge is organized, retrieved, used in decision-making, and modified in light of experience. Components of this process include pathways for transmission and processing within organizations and characteristics of storage and retrieval systems.

Also part of the organizational memory process is "forgetting"—conceived of variously as knowledge depreciation (Argote & Epple, 1990) or intentional "unlearning" (Hedberg, 1981; Nystrom & Starbuck, 1984). Levitt and March (1988) suggest that craft-based organizations and high-level managers tend to rely on tacit and ambiguous information, more so than do bureaucrats and lower-level managers, and that the extent to which knowledge is tacit will have direct implications for active memory decay. The efficacy of storage and retrieval systems is generally regarded as a key dimension in explaining variation in organizational learning. Nystrom and Starbuck (1984) portray the role of top managers as being critical to an organization's ability to unlearn and therefore its ability to learn. According to these authors, executive leaders have a dominating effect on organizational learning processes. So their cognitive structures are very important since they can block recovery from crisis. "Organizations succumb to crises largely because their top managers, bolstered by recollections of past successes, live in worlds circumscribed by their cognitive structures. Top managers misperceive events and rationalize their organization's failures" (p. 58). Hedberg (1981) agrees that organizations must learn by discarding misleading information. This is particularly essential to organizational restructuring and reorientation that is rooted in double-loop learning.

Knowledge management

Also contributing to organizational memory are systems for the management of knowledge and information, including diffusion and transmission mechanisms, and storage and retrieval processes. Much like Huber (1991), Tiler and Gibbons (1991) identify four kinds of knowledge-intensive information activities within organizations: innovation (creation), transfer, allocation (filtering), and coordination. The quality of these activities directly depends on the ways in which information is encoded, pathways for transmission, processing within the organization, and characteristics of storage and retrieval. But several theorists have concerned themselves with *how* information is diffused. For example, Brown and Duguid (1991) suggest that narration (story-telling), collaboration, and social construction are three central features of work practice, yet training activities are mostly concerned with the transmission of information. They propose that learning needs to be understood in terms of communities being formed within the organization. Data from a variety of sources reveal how inadequate communication networks can intrude on organizational learning processes.

The role of the medium of communication in the transmission and diffusion of information within and between organizations has caught the attention of several researchers. Eisenstat (1985), for example, considered knowledge transfer strategies. She proposed that the written word and conversation are both forms of non-behavioral information transfer juxtaposed to, as an illustration, on-the-job training including observation, personal experience, and feedback on performance. Daft and Huber (1987), in considering a variety of modes of information transfer, concluded that "new media are valuable for equivocality reduction to the extent that they increase feedback and encourage a jointly constructed interpretation among individuals" (p. 24).

Finally, methods for locating, storing and retrieving information are essential processes to organizational memory (Huber, 1991). Hedberg (1981) suggests that information management systems define the scope, currency, and accuracy of information within the organization. And while there is some support for the use of information technology to bolster organizational memory (e.g. Hughes, 1991), Levitt and March (1988) caution that, although automation makes retrieval more reliable, it is not necessarily a good thing, since it places limits on the richness of data, validity checks, and the like.

Organizational knowledge development

Experiencing as knowledge creation and acquisition

All organizations begin with some knowledge ("congenital"), but after "birth" knowledge is acquired or created through experiences (Huber, 1991) and organizational learning is closely linked with the experiences encountered by the organization (Garvin, 1993; Nonaka, 1991; Srivastva, 1983). Experiences take many shapes and forms: they may be real or simulated, planned or incidental, retrospective (reflections on past events) or concurrent (reflection on current events). Much theoretical work on organizational learning has concerned itself with activities and processes that are considered planned learning experiences.

In situating their argument for intentional unlearning as a viable route to organizational learning, Nystrom and Starbuck (1984) implore top managers to reconsider their beliefs and practices through continual experimenting; expressing a willingness to deviate from practices they consider optimal in order to test assumptions. Continual experimentation is manifest in a variety of forms (Garvin, 1993; Huber, 1991), most of which embody some sense of trial-and-error learning. Organizations closely observe the effects of their actions on the environment and conjecture about cause and effect relationships.

Huber (1991) acknowledges the role of program evaluation [*see Chapter 7— Eds.*] and action research [*see Chapter 9—Eds.*] in systematizing the trial-and-error process. Designed to enhance effectiveness and utility, "participatory" models advocate a role for organization members in implementing systematic inquiry. Argyris and Schön (1991) and Comfort (1985) argue that participatory action research, a process where organization members are both subjects and co-researchers, will generate internal commitment to the results of inquiry and "create opportunities for individual participants to engage in the re-examination of their basic assumptions so that they may invent new meanings more congruent to actual conditions" (Comfort, 1985, p. 105). Whereas the concept of participation is viewed as integral to learning from systematic inquiry, also critical is the capacity for reflection and self-appraisal (Frey, 1990; Huber, 1991).

Three other forms of experiencing are thought to give rise to organizational learning. One of these, vicarious learning or the development of corporate intelligence through the experiencing of other organizations (Garvin, 1993; Huber, 1991), can occur in a variety of ways. These include the transfer of productivity gains across organizations or products (Argote & Epple, 1990) and the diffusion of new practices throughout a partner organization through personnel rotation (Cohen, 1991). Second, learning from others can be imitative or mimetic; organizations learn through acquiring, encoding, assimilating, and storing for future use the behaviors of others (Leithwood & Aitken, 1995; Levitt & March, 1988). Third, organizations can experience vividly through the use of simulation and games (De Geus [1988] suggests that the real issue is not necessarily whether an organization will learn from gaming or not but whether it will learn sufficiently quickly to benefit from the new knowledge).

Despite strong advocacy for experiencing, some cautions are evident in the literature regarding the extent to which organizations ought to rely on experiencing or experimenting as a means of enhancing learning. According to Dery (1986), stress on experimental learning is a deficiency to the extent that the lack of use of data arising from the experiment is most often assumed to be a crisis or problem to be solved. Another concern arises over the absence of strong and visible links between experimentation and organizational performance. Nevertheless, Levitt and March (1988) maintain that "success is the enemy of experimentation" (p. 334) and recommend that organizations strive to adopt a culture of experimentation.

Acquiring knowledge from the environment

Searching, noticing, and discovering activities are vital determinants of an organization's ability to acquire knowledge from its environment. Kiesler and Sproull (1982) distinguish between problem-solving and problem-sensing processes within the organization. Problem-sensing is reflected in the cognitive processes of noticing and constructing meaning about environmental change so that organizations can take action. They see managerial problem-sensing as a necessary precondition for organizational adaptation and learning.

Three processes—scanning, focused search, and monitoring—are search strategies routinely implemented by organizations according to Huber (1991). Logistic systems are designed to foster aggressive search routines, whereas interpretive systems such as face-to-face contact and group decision support systems are devised to make sense of input from the environment (Daft & Huber, 1987; Louis, 1994). Hedberg (1981) suggests that such perceptual filters can be made conducive to environmental search and discovery.

While search, notice, and discovery activities provide environmental input into organizations, member turnover, succession or *grafting* are also mechanisms for acquiring knowledge (Huber, 1991; March, 1991; Simon, 1991). However, there is little agreement about the merits of turnover for organizational learning. Some theorists (e.g. Virany et al., 1992) suggest that organizational turnover is conducive to double-loop learning to the extent that it eliminates the need for significant "unlearning" to take place. According to Simon (1991), on the other hand, turnover can be a barrier to innovation because it increases training and socialization costs. However, he does recognize that change in orientation implies fundamental change, and that it is generally cheaper and quicker to import new expertise and dismiss the old than to engage in massive re-education. Nystrom and Starbuck (1984), on the contrary, suggest that the eradication and replacement of top managers is not always an attractive solution and that other unlearning alternatives may be more compelling.

Many studies have focused on organizational responses to environmental change (e.g. Levin, 1993; Levin & Ezeife, 1994) and have underscored the importance of "triggering events" (e.g. environmental turbulence) in stimulating such responses. Fiol and Lyles (1985) conjecture that organizational learning requires both stability and change. Whereas they acknowledge the triggering role of turbulence, stability is essential to the reduction in information demand and cognitive workload, allowing for the assimilation of new patterns and beliefs.

Dysfunctional learning habits

Having considered how information and knowledge are generated, acquired, and managed by organizations, we briefly consider organizational processes and effects that get in the way of learning. These are variously referred to as dysfunctional learning habits (Louis, 1994) or organizational learning disabilities (Senge, 1990b). Two dimensions of dysfunctional behaviours have received significant attention in the literature. The first is the concept of *competency traps* (Levinthal, 1991; Levitt & March, 1988) or problematic organizational routines that are well established, reinforced, and difficult to recognize and ameliorate. Competency traps occur when "favourable performance with an inferior procedure leads an organization to accumulate more experience with it, thus keeping experience with a superior procedure inadequate to make it rewarding to use" (Levitt & March, 1988, p. 322). While awareness of alternative superior procedures is obviously essential to break competency traps, a culture of ongoing experimentation is likely to be a potent (preventative) remedy.

A second and related aspect of dysfunctional habits are *organizational defensive routines*.[7] Argyris (1990) identified three patterns of organizational defense:

---

7 Argyris, 1990, 1993; Argyris & Schon, 1978, 1991; Nystrom & Starbuck, 1984.

1. A person accuses an inquirer of being too rational or too scientific when the inquirer asks a question about testing hypotheses or choosing among claims of causality.
2. A person believes rationality and logic are relevant, but should not be pushed when someone is upset or highly emotional.
3. A person accuses the inquirer of being "too judgmental" or "too evaluative" when the inquirer questions the validity and appropriateness of a particular intervention.

These defenses endanger the credibility of organizational learning and are known as "defensive reasoning" techniques. Premises, inferences and conclusions are either tacit or, if explicit, subjected to no other tests except those involving the personal biases of the creator.

Implications for fostering organizational learning in schools

Research and theory about organizational learning continues to accumulate at a rapid pace outside of the field of education. Recent work in education—the present volume being no exception—underscores the promise of organizational learning as a line of inquiry for understanding and improving the administration of educational organizations. This section offers a number of guidelines to educational leaders who wish to further develop the organizational learning capacities of their schools.

Guideline 1: Install or improve existing mechanisms for environmental noticing, searching and discovering

Educational organizations need to take steps to change their posture toward environmental uncertainty. They need to understand their environments more fully in terms of cause and effect relationships and they need to make appropriate anticipatory organizational changes. In short, educational organizations should be less inclined to adapt to change in the environment and be more interpretive, anticipatory, and proactive. At the district level the establishment or maintenance of a strong research department/office holds much promise. A subcommittee reporting to the academic council or cabinet might carry out similar functions at the school level. Regardless of the nature of the structures installed, mechanisms for interpreting incoming environmental information will be necessary.

Guideline 2: Construct and support interpretive systems and opportunities for dense interpersonal exchange

Social exchange and interaction is vital to organizational learning and the propensity for learning will be a direct function of the number of opportunities for social interaction. These might include program evaluation teams, pilot program implementation teams, writing teams, policy analysis teams, and the like.

These structures should be inclusive and, in the ideal, should provide interpretive mechanisms for knowledge generated locally or acquired through environmental scanning.

Guideline 3: Seek, create, and monitor "experiencing" opportunities

Personnel exchanges with other school districts, borrowing and trying other organizations' policies, and imitating other organizations perceived to be successful are examples of organizational experiences that would serve to enhance organizational learning, assuming they are carefully monitored, interpreted, and shared. As March (1991) notes, organizations must strike a balance between experimenting (with longer time frames, less certainty) and exploitation (more concrete, immediate). Levin (1993) supports the emphasis on experimenting by arguing that educational organizations need to get away from their day-to-day activities and responsibilities. But learning from exploitation is possible too.

Guideline 4: Balance local human resource development with grafting

Although research is not clear on this issue, acquisition of external knowledge and expertise through recruitment appears to be a promising strategy for educational organizations. It will be necessary, nevertheless, to strike an appropriate balance (defined by local parameters) between the recruitment of external personnel and personnel developed locally. Lane and Murphy (1988) recommend redesigning staff acquisition processes as a tool for building school cultures. The rapid turnover in the principalship, at present, in many school systems offers a rare opportunity to act on this guideline.

Guideline 5: Revise leadership training and socialization opportunities

Principles of transformational leadership (Leithwood, 1992) are very much in line with an approach to leadership that is likely to engender organizational learning capacity in educational organizations. The foregoing review suggests, as prudent leadership development and socialization, activities that centre on the following skills: being able to recognize the fundamentality of change and the importance of differentiating espoused organizational theories from theories-in-use; being able to surface in oneself and in staff current theories-in-use; being able to think systemically and to recognize that cause and effect are often separated in time and space; being able to recognize and eliminate or reduce organizational defensive routines and dysfunctional learning habits; and, finally, being able to adopt an interpretive posture that values the importance of understanding context over the blind implementation of methods and techniques.

Guideline 6: Install or improve existing district performance indicator systems

Perhaps the most striking difference between public sector organizations and business and industry has to do with the measurability, if not clarity, of goals. This is not to say that viable district monitoring systems cannot be constructed and installed. Leithwood and Aitken (1995) provide an excellent example. These authors make the claim that a district monitoring system can provide information regarding the organization's state of health and regularly collected information that can be translated into courses of action informed by strategic planning. As with mechanisms to scan, search, and discover from the environment, a performance indicator system, in order to be truly effective, requires interpretive learning systems to translate and deliberate about the meaning of data.

Guideline 7: Revisit, reinterpret and revise organizational routines, policies, and symbolic entities

Existing policies, procedures, and organizational artifacts can be valuable sources of theories-in-use. These are the external media within which organizational knowledge resides and the organizational code is defined. Engaging in serious reflective efforts to analyze these routines for congruency with espoused theory can be a useful way to explore tacit, hidden, and underlying assumptions that are held but perhaps not even recognized. It is when a learning system is able to accomplish this difficult task that double-loop learning becomes possible.

Guideline 8: Systematize and automate information storage and retrieval systems

Another problem for schools and school districts trying to become learning organizations has to do with the forms used for transmitting, communicating, storing, and retrieving information. The norm in schools is verbal communications (Leithwood & Dart, 1996) which poses enormous problems for the development and effective use of storage and retrieval systems. It will be prudent to work toward systematizing and automating data storage and retrieval systems such that valuable organizational knowledge can be readily captured, stored, analyzed, and recovered. Such a system will require demonstrated utility and ongoing training for personnel at different organizational roles and levels.

These eight guidelines have not been listed in order of priority. Each will present, obstacles, roadblocks and hurdles as educational leaders work to implement them locally. As with any planned change strategy, it will be necessary to develop explicit local objectives for each guideline and to take stock of current practice. Strategies suited to the local context can then be developed to ameliorate obstacles that are encountered.

Conclusion

Knowledge of organizational learning has developed mainly in the context of organizations that do not look at all like schools or school districts. Nevertheless, an organizational learning perspective offers promising directions for research, school redesign, and the renewal of leadership. In particular, the installation of apparently highly rational and mechanistic logistic systems coupled with relatively open and free-wheeling interpretive learning systems would appear to fit very nicely within the current reform agenda.

Organizations that learn are extraordinarily open, thrive on experimentation and risk, and tolerate ambiguity. At the same time such organizations are able to construct consensual interpretations, as well as surface and eliminate hidden barriers to collective learning. Few schools or school districts currently bear much resemblance to this profile. But it seems clear that organizational change and development in this direction can only enhance the reform effort.

References

Argote, L. & Epple, D. (1990). Learning curves in manufactoring. *Science* 247: 920–924.

Argyris, C. (1990). Inappropriate defenses against the monitoring of organization development practice. *The Journal of Applied Behavioral Science* 26 (3): 299–312.

Argyris, C. (1993). *Knowledge for action: A guide to overcoming barriers to organizational change.* San Francisco, CA: Jossey-Bass.

Argyris, C. & Schön, D.A. (1978). *Organizational learning: A theory of action perspective.* Reading, MA: Addison-Wesley.

Argyris, C. & Schön, D.A. (1991). Participatory action research and action science: A commentary. In W.F. Whyte (ed.), *Participatory action research* (pp. 85–96). Newbury Park, CA: Sage Publications.

Attewell, P. (1992). Technology diffusion and organizational learning: The case of business computing. *Organization Science* 3 (1): 1–19.

Bandura, A. (1977). *Social learning theory.* Englewood Cliffs, NJ: Prentice-Hall.

Bandura, A. (1986). *Social foundations of thought and action: A social cognitive theory.* Englewood Cliffs, NJ: Prentice-Hall.

Brown, J.S. & Duguid, P. (1991). Organizational learning and communities-of-practice: Toward a unified view of working, learning, and innovation. *Organization Science* 2 (1): 40–57.

Cohen, M.D. (1991). Individual learning and organizational routine: Emerging connections. *Organization Science* 2 (1): 135–139.

Comfort, L.K. (1985). Action research: A model for organizational learning. *Journal of Policy Analysis and Management* 5 (1): 100–118.

Cousins, J.B. (1996). Understanding organizational learning for educational leadership and school reform. In K.A. Leithwood (ed.), *International handbook of educational leadership and administration.* Boston: Kluwer Academic Publishers.

Cousins, J.B. & Earl, L.M. (1992). The case for participatory evaluation. *Educational Evaluation and Policy Analysis* 14 (4): 397–418.

Cousins, J.B. & Earl, L.M. (eds.). (1995). *Participatory evaluation in education: Studies in evaluation use and organizational learning.* London: Falmer Press.

Cousins, J.B. & Leithwood, K.A. (1993). Enhancing knowledge utilization as a strategy for school improvement. *Knowledge: Creation, Diffusion, Utilization* 14 (3): 305–333.

Daft, R.L. & Huber, G.P. (1987). How organizations learn: A communication framework. *Research in the Sociology of Organizations* 5: 1–36.

Daft, R.L. & Weick, K.E. (1984). Toward a model of organizations as interpretation systems. *Academy of Management Review* 9 (2): 284–295.

De Geus, A.P. (1988). Planning as learning. *Harvard Business Review*: 70–74.

Dery, D. (1986). Knowledge and organizations. *Policy Studies Review* 6 (1): 14–25.

Eisenstat, R.A. (1985). *Organizational learning in the creation of an industrial setting*. Unpublished doctoral dissertation, Yale University, New Haven, CT. (DAI 46/12B, p. 4438).

Fiol, C.M. & Lyles, M.A. (1985). Organizational learning. *Academy of Management Review* 10: 803–813.

Frey, K. (1990). Strategic planning: A process for stimulating organizational learning and change. *Organizational Development Journal* 8: 74–81.

Fullan, M.G. (1993). *Change forces: Probing the depths of educational reform*. London: Falmer Press.

Garvin, D.A. (1993). Building a learning organization. *Harvard Business Review* July–August, 78–91.

Hedberg, B. (1981). How organizations learn and unlearn. In P.C. Nystrom & W.H. Starbuck (eds.), *Handbook of organizational design, volume. 1: Adapting organizations to their environments*. New York: Oxford University Press.

Huber, G.P. (1991). Organizational learning: The contributing processes and the literature. *Organization Science* 2 (1): 88–115.

Hughes, T.S. (1991). *Organizational learning in rural development agencies*. Unpublished doctoral dissertation, Cornell University, Ithica, NY. (DAI 52/06A, p. 1985).

Kiesler, S. & Sproull, L. (1982). Managerial response to changing environments: Perspectives on problem sensing from social cognition. *Administrative Science Quarterly* 27: 548–570.

Lane, B.A. & Murphy, J. (1988). Building effective school cultures through personnel functions: Staff acquisition processes. *Journal of Personnel Evaluation in Education* 2: 271–287.

Lant, T.K. & Mezias, S.J. (1992). An organizational learning model of convergence and reorientation. *Organization Science* 3 (1): 47–71.

Leithwood, K.A. (1992). The move toward transformational leadership. *Educational Leadership* 49 (5): 8–12.

Leithwood, K.A. & Aitken, R. (1995). *Making schools smarter*. Thousand Oaks, CA: Corwin.

Leithwood, K.A. & Dart, B. (1996). Commitment-building approaches to school restructuring: Lessons from a longitudunal study. *Journal of Education Policy* 11 (3): 377–398.

Levin, B. (1993). School response to a changing environment. *Journal of Educational Adminstration* 31 (2): 4–21.

Levin, B. & Ezeife, A. (1994). *Schools and school systems coping with a changing world*. Paper presented at the annual meeting of the Canadian Society for the Study of Education, Calgary AB.

Levinthal, D.A. (1991). Organizational adaptation and environmental selection—interrelated process of change. *Organization Science* 2 (1): 140–145.

Levinthal, D. & March, J.G. (1981). A model of adaptive organizational search. *Journal of Economic Behaviour and Organization* 2: 307–333.

Levitt, B. & March, J.G. (1988). Organizational learning. *Annual Review of Sociology* 14: 319–340.

Louis, K.S. (1994). Beyond bureaucracy: Rethinking how schools change. *School Effectiveness and School Improvement* 5 (1): 2–24.

Lovell, R.D. & Turner, B.M. (1988). Organizational learning, bureaucratic control, preservation of form: Addition to our basic understanding of research utilization in public organizations. *Knowledge: Creation, Diffusion, Utilization* 9 (3) 404–425.

Lundberg, C.C. (1989). On organizational learning: Implications and opportunities for expanding organizational development. *Research in Organizational Change and Development* 3: 61–82.

March, J.G. (1991). Exploration and exploitation in organizational learning. *Organization Science,* 2 (1): 71–87.

Nonaka, I. (1991). The knowledge-creating company. *Harvard Business Review* November: 96–104.

Nystrom, P.C. & Starbuck, W.H. (1984). To avoid organizational crises, unlearn. *Organizational Dynamics* 12: 53–65.

Saffold, G.S. (1988). Culture traits, strength, and organizational performance: Moving beyond "strong" culture. *Academy of Management Review,* 13 (4): 546–557.

Senge, P.M. (1990a). The learner's new work: Building learning organizations. *Sloan Management Review*: 7–23.

Senge, P. M. (1990b). *The fifth discipline: The art and practice of organizational learning.* New York: Doubleday.

Shujaa, M.J. & Richards, C.E. (1989). Designing state accountability systems to improve school-based organizational learning. *Administrator's Notebook* 33 (2): 1–4.

Simon, H.A. (1991). Bounded rationality and organizational learning. *Organization Science* 2 (1): 125–134.

Srivastva, P. (1983). A typology of organizational learning systems. *Journal of Management Studies* 20 (1): 7–28.

Sterman, J.D. (1989). Modeling managerial behavior: Misperceptions of feedback in a dynamic decision-making experiment. *Management Science* 35 (3): 321–339.

Taylor, B. O. (1986). *Metasensemaking: How the effective elementary principal accomplishes school improvement.* ERIC Document Reproduction Service (No. ED 278 123).

Tiler, C. & Gibbons, M. (1991). A case study of organizational learning: The UK teaching company scheme. *Industry and Higher Education* 5 (1): 47–55.

Tushman, M.L. & Romanelli, E. (1985). Organizational evolution: A metamorphosis model of convergence and reorientation. *Research in Organizational Behaviour* 7: 171–222.

Virany, B., Tushman, M.L. & Romanelli, E. (1992). Executive succession and organization outcomes in turbulent environments: An organization learning approach. *Organization Science* 3 (1): 72–91.

Whyte, W.F. (1991). Introduction. In W. F. Whyte (ed.), *Participatory action research* (pp. 7–15). Newbury Park, CA: Sage Publications.

Wills, F.G. & Peterson, K.D. (1992). External pressures for reform and strategy formation at the district level: Superintendents' interpretations of state demands. *Educational Evaluation and Policy Analysis* 14 (3): 241–260.

Woods, P. (1993). Responding to the consumer: Parental choice and school effectiveness. *School Effectiveness and School Improvement* 4 (3): 205–229.

11

School Development and Organizational Learning: Toward an Integrative Theory

Janna C. Voogt, Nijs A.J. Lagerweij, and Karen Seashore Louis

Introduction

Schools change constantly. Demands for improvement or renewal from local districts and states or provinces provides one impulse for change, and schools also develop their own initiatives. In addition to these systematic change activities one can also find autonomic development processes as well. Like people, schools age. As they do, school cultures (habits, norms, and values) become more embedded and the institution evolves toward the next stage (Schein, 1992). Both this naturally evolving organizational life and the deliberately initiated changes are influenced by a third change force: "normal crises" (Louis & Miles, 1990) or synchronic events. These include unpredictable but common calamities—illness, mergers, sudden demographic shifts, unanticipated personnel changes, and natural disasters—and also events that provide opportunities for revitalization. To understand the process of change in schools—and organizational learning—one needs to attend to all of these components of what we call *school development*.

In this chapter we argue that most theories and research on change processes concern the paradigm of deliberate or planned change. In our view, this singular focus limits our understanding of how schools change: a broader perspective is a necessary condition for genuine organizational learning. To define that broader view we briefly examine the planned change research traditions that influence our perspective on total school development and organizational learning. We then go on to examine how each of these traditions accounts for organizational learning. Finally, we turn to a discussion about how the idea of school development contributes to the concept of organizational learning.

Research on planned change

Among the research traditions that are particularly relevant to our argument are organization development, effective schools, school improvement, and systems analysis. We focus on these because they include traditions of interventions for planned change in addition to theory.

Organization development
The concept of school development has its origins in the research and support tradition of "OD" or organization development. This approach—derived from the social-psychological studies of group behavior and change (Lewin, 1947) and described in several reviews (Fullan et al., 1980; Hopkins, 1982; Schmuck et al., 1977; Schmuck & Runkel, 1985)—is characterized by its focus on *group dynamics,* not the skills of its individual members (Schmuck, et al., 1977, p. 3). The main intervention in the classic OD tradition is to create conditions for cognitive and affective change using *self-reflective, self-correcting* methods based on *data feedback* from a trained internal or external OD consultant. The OD approach was highly popular during the 1970s in centers that facilitated change processes in schools. Although initially developed in the U.S., it has had a significant impact outside the Anglo-American publications (Scheerens & Creemers, 1989; Vandenberghe, 1987; Voogt, 1986).

Effective schools
The 1960s and '70s were periods of significant public investment in education, particularly in programs (curricula and supplementary services) focused on improving the performance of poor and minority students. By the 1980s, policymakers increasingly questioned whether this investment paid off in pupils' outcomes. At the same time, Edmonds (1979) pioneered an alternative question: are there schools that are uniquely effective for children from poor families? The idea of "the effective school" (ES) focused on buildings with higher pupil achievement on standardized reading and arithmetic tests than other schools with similar student populations.

Early research in the ES tradition pointed to five common factors that make a difference between more and less adequate school settings:

1. A school leader's focus on the learning process
2. A strong accent on learning basics
3. A safe and orderly climate
4. High expectations of pupil outcomes
5. Frequent monitoring and evaluation of pupil outcomes

Based on these five factors, School Improvement Programs (SIPs) that linked OD-based interventions with the ES findings were developed and implemented during the 1980s (Kyle, 1985; Louis & Miles, 1990; Purkey & Smith, 1983).

The ES research tradition also influenced educational research in other countries although (except in the United Kingdom) it has had limited impact on national school improvement initiatives. For example, in the Netherlands two university institutes became centers of effective schools research and sharpened debates about methods and variables to set educational quality (Creemers, 1987; Scheerens & Creemers, 1989). Most large-scale Dutch and U.S. research has been carried out with elementary schools. However, in the United Kingdom, the effective schools research strategy has included more extensive attention to secondary schools as well (Thomas et al., 1995; Willms & Raudenbusch, 1989). Effective schools studies have become increasingly methodologically sophisticated, using similar instruments in different national contexts, and increasingly using large-sample, longitudinal databases and hierarchical analytic techniques. However, programmatic linkages with the OD tradition have been limited in the European context.

School improvement

In 1982, the Center for Educational Research and Innovation (CERI), as part of the Organization for Economic Cooperation and Development (OECD), initiated a project to develop internationally applicable theory and practice related to educational policy and improvement. The International School Improvement Project (ISIP) involved 14 countries and lasted from 1982 to 1986. Six books, six technical reports, and many articles were published during this project, among which was its theoretical framework (Van Velzen et al., 1985).

To ensure a systematic analysis, ISIP worked through a series of cross-national groups focused on specific topics:

Area 1   School-based review for self-improvement.
Area 2   Principals and international change agents in the school improvement process.
Area 3   The role of support in school improvement processes.
Area 4   Research and evaluation in school improvement.
Area 5   Development and implementation of school improvement policies
Area 6   Conceptual mapping of school improvement.

The ISIP idea of *school improvement* drew largely on the OD tradition. The participants were, however, willing to "peer over the fence" and address pupil outcomes as well as change processes. The ISIP definition of school improvement reflects this hybrid perspective: "a systematic, sustained effort aimed at change in learning conditions and other related internal conditions in one or more schools, with the ultimate aim of accomplishing educational goals more effectively" (Van Velzen et al., p. 48).

Using this definition, school improvement in ISIP broadened the older OD traditions, yet it did not fully integrate the evolving effective schools approach.

The ISIP approach continued to emphasize internal conditions, such as school organization, management, or decision-making. The ES stream, in contrast, maintained its emphasis on educational outcomes as assessed by national or standardized tests, and classroom processes. Clark, Lotto & Astuto (1984) noted this continuing difference, and scholars from both traditions have persisted, with modest success, in their search for a common ground for theory and intervention (Reynolds et al., 1996). The development of a new was journal devoted to both streams (*School Effectiveness and School Improvement*) has assisted in the efforts to look for unifying findings and themes.

Contingency approaches[1]

A way of distinguishing among models of organizational analysis is between *convergence perspectives,* where more productive organizations in any setting will resemble each other, and those emphasizing *organizational pluralism,* or sub-units pursuing their own aims and interests (Owens, 1981). All of the models discussed above can be considered as belonging to the first category.

Within the pluralistic perspective, the organization is regarded as a quasi-autonomous, self-regulating system. Organizations are viewed as open systems that strive for survival, growth, and exchange of materials and resources with their environments. Schools, as open systems, interact with their settings in ways that are informally (or formally) regulated by a constant feedback cycle. This contingency principle requires a match between the organizational structure and its external relationships (Mintzberg, 1979) or between organizational structure and its primary technologies (Peterson et al., 1996), but does not imply a single model of effective design.

In this view a system evolves through two processes: differentiation and integration (Lawrence & Lorsch, 1967). Differentiation occurs through division of labor and professionalization. In schools, differentiation—e.g., the growth of subject matter and pupil service specializations—mirrors the teaching–learning process, or the primary process in schools (Allegro, 1973). Integration (coordinating functions that have been divided) occurs through the regulation of hierarchical and lateral communication, and the encouragement of informal communication. Allegro (1973) calls these the secondary or organizing processes of schools. These two domains of integration and differentiation yield two axes that shape the developmental field for schools:

- The educational process moves from uniformity toward differentiation—this development parallels the change of the school's focus exclusively internal to internal and external (Quinn & Rohrbaugh, 1983).
- The organizational structure shifts from control toward flexibility—this development parallels the substitution of organic coordination for formal or rule-based coordination (Morgan, 1986).

---

1 This label covers organizational theories like Human Relations, Management Process Theory, Revisionism, Decision-Making Theory, Communication and Information Theory, System Theories, Rational Goal-Based Thinking and Contingency Theories.

In the Netherlands, school diagnoses that cover both the instructional (primary) and organizational (secondary) domains of school organization are influential (Caluwé et al., 1988; Marx, 1975; Voogt, 1989). In the American situation, recent "restructuring" approaches take both domains into account (Murphy, 1991; Newmann & Wehlege, 1995). In order to "make schools smarter"—that is, to realize contingency principles—systems for monitoring school progress in both domains of school functioning and the relationship with school context are taken into account (Leithwood & Aitken, 1995).

One weakness of the contingency approach (noted also in Chapter 1) is the absence of empirical evidence that more than a modest number of organizational designs are feasible or functional. While arguing for an open systems and contingency perspective, most of the authors mentioned—Marx (1975), Mintzberg (1979), and Quinn and Rohrbaugh (1983)—have proposed developmental models based on the two axes. The premise of this distinction is that the developmental stage on one axis should match the developmental stage on the other axis. The basic assumption in relationship to these models is that schools can develop from one (more simple) stage to the next (increasingly complex). As Marx (1987) concluded, this development moves "from segmentation through a collegial organizational model to an innovative organization" (see also Schein, 1992).

Empirically this is, of course, not often the case in schools: schools do not usually adjust their organizational structure to the changing educational process (again, see Chapter 1). For example, more flexible and supportive organizational structures at the district/provincial/state level do not accompany a school's desire to initiate broader educational content and more varied instructional strategies. "Site-based management" is seen as compensating for this inconsistency (Caldwell & Spinks, 1988), but there is limited research evidence to suggest that this structural change typically leads to more productive schools.

According to Morgan (1986), one can "read" an organization's developmental stage based on analyzing its "artifacts" (Schein, 1992) or the *structural determinants* and its *dynamics*. Lagerweij & Voogt (1997) conclude that they can cluster structural determinants into seven factors:

1. The school's innovation policy
2. The individual team members
3. School as an organization (team functioning, climate, culture, etc.)
4. Educational content and the teaching–learning process
5. School leadership and internal/external support
6. Material and other preconditions
7. Outcomes

We could argue that these determinants are the structural components of the (school development) system. And, we also suggest that they are equally significant in determining organizational learning.

Organizations are dynamic due to the complex interactions between the determinants; the nature of this interaction is elusive in both research and daily practice. Lagerweij and Voogt (1990), Marks and Louis (1997) and Bryk, Camburn and Louis (1996), among others, focus on "innovation capacities" at school or classroom level (predispositions, values, and skills) associated with the determinants. However, causal connections between changing capacities in one determinant and "improvements" in another are not well documented in larger scale studies, in part because of the limited number of longitudinal multi-site/multi-level studies. In practice there are always surprising developments that are best explained by ad hoc decisions, unintended outcomes, or unanticipated events (March & Olsen, 1976).

Since we argue that organizational learning is a major factor in school development and particularly in school dynamics, we will look in the next section at the learning processes that are inherent in the approaches described above.

Organizational learning within these contexts

The balanced development of schools as organizations theoretically requires modest internal consistency among the structural elements, the culture, and the organizational goals. On the other hand, stage theories suggest that the configuration of these elements changes at every developmental level. Organizational analysts increasingly observe that the organizational learning process varies between developmental levels, each of which has different goals and a different character (Argyris, 1990; Swieringa & Wierdsma, 1990; Tjepkema, 1994). The learning principle inherent in contingency theories requires effective organizations to deal with their specific environment and technologies (Fiol & Lyles, 1985) through 1) adaptation, 2) sharing assumptions, 3) developing knowledge about relationships between actions and outcomes, and 4) institutionalized experience (Tjepkema, 1993). Yet, as with humans, organizational growth and development is often inconsistent. Adolescent feet often grow before a beard appears and the appearance of gray hair does not predictably occur before or after we need reading glasses. In other words, development is erratic and occasionally incongruous.

At the same time, learning processes differ in character or complexity. Argyris and Schön's (1978) distinction between *single-loop,* and *double-loop* learning is well known, and refers to adjustments in behavior or challenges to fundamental assumptions that result from environmental feedback. Senge (1990) adds a third dimension, *practice,* which refers, in turn, to the principle of team learning: "a group of talented individual learners will not necessarily produce a learning team, any more than a group of talented athletes will produce a great sports team" (p. 257).

Likewise Swieringa and Wierdsma (1990, pp. 39–48) link the idea of learning loops to domains of learning and at the same time to categories of learning and to labels of learning outcomes. In doing so, they make the same distinction that we have drawn between improvement, systemic change, and development, albeit in the different context of business organizations. Their notion of development does not, however, fully incorporate an analysis of the unpredictable

Table 11.1 Categories of organizational learning and its outcomes.

Learning loop	Learning domain	Learning category	Learning outcome
single	rules	must, have to, can, allowed	improvement
double	insights	know, understand	innovation
triple	principles	dare/want, & to be	development

nature of environmental learning (we will return to this issue later). With this conceptual framework we now can typify the organizational learning processes in the distinctive research traditions identified above.

In chronological order we began with the OD tradition of the 1970s, which tried to find a solution for the segmentation problem and aimed for flexibility in both the organizational and educational domain. The implicit organizational learning process advocated by this approach is characterized by *reflexive self-analytic methods* (data feedback) and concerns not only matters of rules and behavior, but also *why questions*: "Why are we doing this in this way?" and even more often "Why can't we do it another way?" This implies that culture becomes articulated and discussible by "examining the artifacts" or "sharing assumptions."

Learning strategies include systematic examinations of the teaching–learning process by reconstructing and discussing it at team meetings to develop knowledge about relationships between actions and (process) outcomes. Clearly these are characteristics of the double-loop learning process that depends on questioning the relevance of operating norms (Morgan, 1997, p. 88). Schools that can handle this process successfully will often have an organizational model that is collegial (Marx, 1987) with shared decision making, or "a professional bureaucracy" (Mintzberg, 1979) and could be developing toward a matrix model. Hopkins (1995) argues that these schools are "problem-solving schools," and according to Van Wieringen (1993, p. 302) those schools will have a threefold capacity of "self-schooling": 1) the design or policy developing capacity, 2) the operation, implementation capacity, and 3) the evaluating capacity. The organizational learning outcome is innovation.

Chronologically, the next set of interventions that can be linked to learning is the Effective Schools movement. It is no coincidence that this approach parallels (or combines with) a sense of crisis in school performance and the back-to-basics trend in national policies. The emphasis is less on flexibility than on the need for results related to known performance indicators. As in business life (Peters & Waterman, 1982), the emphasis is on the "core business of basic skills," customer-oriented goal formulation, more authoritative leadership and simpler structuring, motivation of staff, and action orientation. In education this became translated into clear learning objectives, focusing on statistically measurable learner outcomes, high expectations, and improving (outcomes) by effective (direct) instruction.

The typical action features associated with the effective school approach are the development of focused educational programs based on the effective schools research, prescriptive training for teacher instruction, and dissemination of publications promoting uniform practice (U.S. Department of Education, 1986). Together these express an action-orientation focused on "How can we do better?" rather than "What should we be doing?" Organizational learning in this context leads to changing of rules and/or behavior and, at least in some cases, may be regarded as single-loop learning that "rests in an ability to detect and correct error in relation to a given set of operating norms" (Morgan, 1986, p. 88). The organizational learning outcome is performance—on test scores, dropout rates, parent satisfaction, etc. As for the organizational structure, the trend is simultaneously toward autonomy *with* accountability, and simplicity (Mintzberg, 1979; Resnick et al., 1995). Proponents of the effective schools models for improvement argue that the dismal performance of many schools in countries with well-funded educational systems necessitates an emphasis on coherence of objectives, classroom practice, and accountability that serves all students.

Movements do not stop when new trends arise: they overlap in time, and are sometimes integrated into the next stage. So, in the next development of the School Improvement approach there are both elements of the international School Improvement/ISIP project, traces of traditional OD approaches, and starting points for new efforts to move forward (Murphy, 1991).

School Improvement Programs (SIPs) emerged around a route aimed at applying school effectiveness findings (ES) plus the management of change (OD). Agreement between principal and teachers on goals, using externally supported planned change programs, conducting audits of current strengths/weaknesses, and formative evaluation of and feedback on progress were characteristic accents within ISIP (Dimmock, 1995). With its emphasis on the guidance of change, the ISIP project was in line with the school-based management (SBM) initiatives (Malen et al., 1990) that dominated the late 1980s and early 1990s.

In turn, the SBM approach soon was succeeded by the broader restructuring efforts of the late 1980s and 1990s (Murphy, 1991; Newmann &Wehlege, 1996). Both approaches (SBM and restructuring) found their causes in the international trends toward decentralization of school policy-making (Louis & Voogt, 1998). Accumulating experience questioned the efficacy of national programs in driving school performance, and many countries sought new routes by simultaneously stimulating school-based management and school-based accountability for measuring and monitoring student outcomes. Many quickly questioned the assumed relationship between restructuring and quality of schooling (Murphy, 1991), although there is some evidence to suggest that it can affect student achievement (Marks & Louis, 1997; Newmann & Wehlege, 1996).

In sum, we have yet to find a serious test of the efficacy of organizational learning in schools as a consequence of approaches based on newer educational research. In particular, there is limited evidence that interventions can be

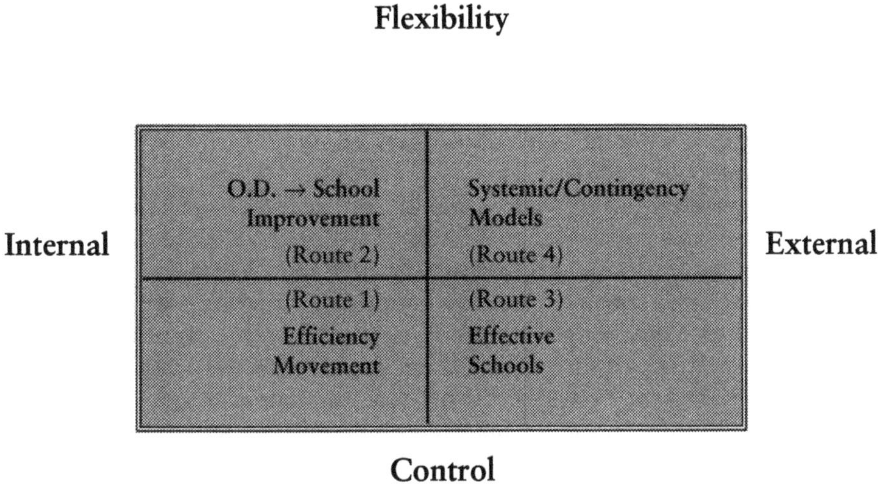

Flexibility

Internal   **External**

O.D. → School Improvement (Route 2)	Systemic/Contingency Models (Route 4)
(Route 1) Efficiency Movement	(Route 3) Effective Schools

Control

Figure 11.1. Planned change/learning traditions.

designed to increase organizational learning beyond those that have already been shown to affect limited single-loop or (at best) double-loop learning.

Figure 11.1 represents the two axes and the related theoretical base and/or intervention strategies. It is interesting to see that developmental perspective the OD approach draws upon the *vertical axis* (the organizational structure—toward flexibility) and the ES approach stresses the *horizontal axis* (the organizational focus—toward external standards). According to Quinn and Rohrbaugh (1983), development goes either first in vertical direction and then horizontal (route 1) or first in horizontal direction and then vertical (route 2). There is, however, no empirical support in educational research for a unidirectional developmental trajectory.

Tracing uncertain theoretical evolution in policy: A case example of "natural contingency?"
No single example can prove a point. However, the problem of overlap between the Effective Schools/Organization Development and the Uniformity/Multiplexity perspectives on school change and learning are well illustrated by a brief history of Dutch educational policy over the past 25 years.

In the 1970s the influential (Socialist) Dutch Minister of Education, Van Kemenade, launched his ideas on a renewal of secondary education that was to lead toward a comprehensive middle school with broad, differentiated, flexible educational content and practice. So the Dutch secondary sector went on their way via "route 1" to "route 2" in Figure 11.1. However, the route was politically difficult, since many politicians and parents outside the Socialist Party[2] were concerned about moving toward a school system whose progress was measured by school-developed criteria rather than by traditional curriculum benchmarks. At this time, national testing in the Netherlands was limited largely to the "school leaving examinations" at the end of secondary school. While these were compulsory, they were high-stakes examinations primarily for those entering higher education.

Shortly thereafter, policy shifted rapidly to "route 3"—effective schools-based policies—which were initiated in primary education in the larger cities. Although the national law still reflected Van Kemenade's vision of the self-directed school, concerns were raised about the gap in performance of children from lower-income and immigrant groups contrasted with native Dutch children—in other words, self-direction did not appear to be well implemented in urban primary schools. Thus, the effective schools approach appeared to offer some research-based promise for immediate solutions for serious reading and language problems. In contrast, the secondary school system (which is not comprehensive but streamed by ability level and work aspirations) was too complex easily to incorporate the ES approach, and many village and private primary schools also largely ignored it.

Now meeting the more comprehensive renewal demands of the 1990s, secondary schools confront a new policy that requires them to move toward a "study house" (a learning community similar to the "route 4" option in Figure 11.1) by continuing "route 2." Yet, while many schools are familiar with the OD focus on group dynamics and teamwork, they have little experience in applying this understanding to their core work—including examining their results and their relationships with the environment of parents, employer organizations, and universities.[3] The government is examining its options to support this development, but the success of this acceleration of the 20-year-old Van Kemenade policies is uncertain.

---

2 Partie van de Arbeid or PvdA. This is one of four major political parties in the Netherlands, and was part of the ruling coalition during the '70s and early '80s, and again during the late '80s and '90s. Even when out of power, it is has a strong influence on educational policy.

3 In the primary sector there are persistent tensions due to inconsistent policy demands (Louis & Voogt, 1997). Primary schools will become more bureaucratic because of the scaling up of school size, which results in mergers and collaboration connections. At the same time they are obliged to use self-reflexive analytical methods for quality assessment (OD approach). In addition, they have to implement inclusive education (broadening educational content, more differentiation) in response to external pressures. This demands a significant reinvention of schooling, drawing on organizational learning frameworks, since there are no models. On the other hand, the inspectorate also evaluates them on standards that originate from the ES approach.

Summary and reflection

Our discussion to date concerns the initiation and implementation of changes in schools that are intended as innovations or improvements. The similarities are:

- The school is the unit of change
- The emphasis is on a planned change approach within the school
- A systemic accountability policy at the district/state/provincial/national level
- Developing school capacities (such as leadership) is seen as a prerequisite

We hear increasing criticism of this theory-of-action from many quarters. Mintzberg (1994) described *The Rise and the Fall of Strategic Planning*; Beer & Eisenstadt (1990) formulated the problem as "Why Change Programs Don't Produce Change." Morgan (1997) asserted, based on his research, that an individual's leverage over work results is limited to the direct sphere of influence of 15%. Along the same lines, one popular solution to increased organizational productivity—Total Quality Management—was shown to fail for 70% of those who implemented it (Beer, 1989) and some excellent companies identified by Peters & Waterman appeared to be in trouble after a three-year follow-up (Easterby-Smith, 1990). Thus, the OD and "planned change" approaches were criticized for their confidence in a linear improvement process and the manageability of organizations (Louis, 1994; Tjepkema, 1994, p. 42).

Perhaps in frustration, or an effort to maintain the myth of human control, a thorough searching for new approaches occurred in the second half of the 1980s. The new movement was called Organization Transformation. In organizational transformation, change processes do not rely on a "tool kit" for internal or external "change agents." Rather, all employees are seen as "tools," especially when the inner human being works in concert with ethical principles. Transformational thinking often links issues and processes of personal identity and development with those of the organization. Transformation thus viewed is a process of continuous individual and group modification punctuated by organization-wide mutations in which the outcome is unsure. The idea of transformation is close to our concept of development because both emphasize periods of discontinuity and of sudden leaps of growth.

In the (business) organizational literature, the term "transformation" is often used as a synonym for the process of *self-development* of organizations. Organizational transformation concerns internally initiated changes, from— collectively and individually—*wanting* the organization to reach certain goals and to grow. Meaningful changes do not occur as a consequence of minor accommodations or external intervention but from more radical changes in the character and culture of the organization (Pedler et al., 1991; Tjepkema, 1994). Sometimes the emphasis is on expanding organizational goals from the immediate (profit or student outcomes) to the longer range (global survival of capitalism or the development of good citizens). In addition, the importance of

organization is refocused on the conditions of the members, such as community and caring (Louis et al., 1996; Sergiovanni, 1994), as well as external performance.

Not surprisingly, in the same period of (inner) human change forces aggre-
gating to the organizational level, (inner) human learning processes also reap-
peared at the organizational level as, for example, in Peter Senge's (1990)
publication *The Fifth Discipline*. The learning organization approach took off
in the managerial consciousness. More recently, publications on organizational
learning appeared in the field of education (Fullan, 1993; Louis, 1994). Still, at
the present time the organizational transformation model remains more an ideal
than a well-implemented model.

Attempts to carry out an organizational transformation approach in education and business have suffered from using old tools and inadequate foundations for policy and learning programs. Another hindrance is the contradiction
between the (internally initiated) organizational transformation perspective and
current reality (change initiatives are frequently initiated from outside the
school). Schools themselves have to choose organizational transformation, and
the role of outsiders is limited to helping schools develop the insight that organizational learning is crucial for school development.

Organizational learning and school development: A conceptual framework

School development, in our view, is a result of the influences outlined above,
namely:

1. An autonomous developmental process (organizational life cycles) includ-
    ing the biggest part of the daily routine
2. Deliberately directed attempts (from within and from outside) to bring
    about educational and organizational changes
3. Major anomalies, and unanticipated events, both positive and negative,
    that must be factored into the organizational learning process

These three influences correspond to the three levels of organizational learning
summarized in Table 11.1, namely single-, double- and triple-loop learning.
They expand on these three levels, however, by incorporating both the 15% of
the improvement "pie" (Figure 11.2) that can be affected by deliberate efforts
to change, and the 85% that is not directly subject to planned intervention. This
leads to the following definition: "School development is an ongoing process in
which the autonomous, coincidental, and deliberately directed changes that
simultaneously affect the functioning of school converge."

The proportions of Figure 11.2 are not empirically based. It assumes that
Morgan's estimate of the potential impact of deliberate change is on the mark,
and also that it is possible to estimate a lower bound of the "unexpected/ran-
dom" component by examining school statistics related to building problems,
absenteeism, turnover, etc. If we suppose that the remaining part is equivalent

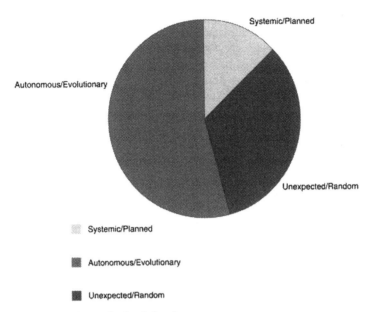

Figure 11.2. Dimensions of school development.

to the autonomous evolutionary processes that dominate the organizational life cycle, then the figure captures at least a heuristic reality. Of course proportions differ between schools, while a major calamity like a fire may overwhelm the planned and life cycle development processes.

What is surprising in studies of organizational learning is the lack of empirical evidence for so much of the development process. We have focused in this review on the planned change research because it is only here that we see a robust knowledge base related to the impact of change on school performance. We also know a great deal about the effects of student demographic characteristics and general social conditions on student achievement from the sociological literature. Because these are always changing, they affect school the evolutionary component of development in ways that are rarely factored in to cross-sectional or short-term longitudinal studies of planned change. Even less is known about the role of the unexpected in the development process—instead, unanticipated events are most often used as post-hoc explanations for outcomes of a planned change activity. In particular, analyses of organizational learning tend to involve research related to planned or strategic efforts to gather and learn from data, although writers such as Daft and Huber (1987) note that unplanned learning is also an additional "tool" to create more effective organizations (see also Smylie et al., 1996). Less attention than is desirable is paid to the indirect effects of "normal crises" on school development, although there are many case studies that testify to its impacts on learning and development (see, for example, Louis & Miles, 1990; Rollow & Bryk, 1995).

Another dominant research stream, on the other hand, identifies the disjuncture between policy and school development (Cohen et al., 1972; Malen et al., 1990; Weiler, 1993). Implicitly we assume that policies are intertwined with both deliberate change effort and the "normal crises" of development. Articulated innovation policies, such as the trend toward decentralization in western countries (Beare & Boyd, 1993), aim at increasing management and/or performance capacities in the school, but also have the effect of dramatically altering relationships and expectations between partners, with unanticipated effects.

In addition to specific innovation policies we can distinguish both ancillary and parallel policies. Ancillary policies aim at supporting innovation policy (e.g., adjusting teacher licensure requirements to correspond to new educational demands). Parallel policies concern educational initiatives that are independent from innovation policy but occur at the same time and interact with centrally stimulated innovation efforts (e.g., changing financial formulas for school funding) (Louis & Voogt, 1997). Both types of policies may also have unanticipated impacts—positive and negative—on the capacities of schools to learn. When such policies are initiated, they may interact with both planned change capacities and may stimulate normal crises; when they become incorporated into daily routine, they become an unacknowledged component of the autonomous developmental process.

Organizational learning as leverage for school development

In this chapter we have argued for a broad view of the development process of schools. This process has three components that we respectively indicated as the systematic/planned dimension, the autonomous/evolutionary dimension, and the unexpected/random dimension. The three dimensions are applicable both to the teaching and learning process (primary process in schools) and to the organizational maintenance and renewal process (secondary process in schools). Together these elements shape a kind of matrix model that captures the concept of school development (Table 11.2). The underlying principles of this model reflect the routine of organizational life as well as the deliberate internal adjustment and consistency of the elements.

In this section we argue that organizational learning is a prerequisite for handling this broad organizational view and putting the dimensions together. When involved groups are able to engage in organizational learning, they will comprehensively understand what school development entails in their own setting. In making this argument, we will view school development through the lenses of Senge's (1990) five disciplines.

Organizational learning means that groups grapple with questions such as: *How* are we doing? *Why* are we doing it this way? And *to what end* are we doing things? The "how question" is about rules and behavior, the "why

Table 11.2. A summary of the elements of school development and organizational learning.

	systematic/planned dimension	autonomous/evolutionary dimension	unexpected/random dimension
primary process			
secondary process			

question" is about culture, norms and values, and the "ends/purposes question" is about identity. In addition, the information that needs to be processed in conjunction with learning must address all of the cells in Table 11.2: issues of organizational life cycles and normal crises must be as much a part of organizational dialogue as the formal planned change processes. Clearly schools will not address all these questions in every cell on every day, nor will learning always involve a conscious attempt to address the questions. Nevertheless, the "triple-loop" learning process cannot occur unless all are considered at some regular intervals.

Team learning
Organizational learning means articulating individual and team learning—not an easy task. Fundamental to this process is *digesting*, that is like "chewing" and "ruminating" learning experiences as a team. As Louis, Kruse and Raywid (1996) note:

> Not all school communities have a strong propensity to learn and change; not all learning organizations support enhanced professionalism. These concepts share a common focus, however, on the need for sustained conversation. (p. 13)

In Senge's (1990) terms:

> The discipline of team learning starts with 'dialogue,' the capacity of members of a team to suspend assumptions and enter into a genuine "thinking together." To the Greeks *dia-logos* meant a free flowing of meaning through the group, allowing the group to discover insights not attainable individually. (p. 10)

Senge stresses the fact that learning-through-dialogue is not a technique, but a discipline that must be mastered and consistently reinforced. Organizations frequently gather information for the ostensible purpose of learning, but then do nothing with it (McDonald, 1995). To self-consciously (although not necessarily according to schedule or plan) process information in a group is really "a shift of mind" (Senge, 1990), in which randomly initiated conversations are taken as serious "learning opportunities" (see Chapter 3).

Systems thinking

The broad view of school development with its matrix of elements shown in Table 11.2 also captures the "systems thinking" that Senge regards as the fundamental "fifth discipline." Without a systemic orientation, including contingency thinking, there is no capacity to integrate information about superficial dissemination activities, environmental conditions, etc. By enhancing each of the other practices needed to support organizational learning, it continually reminds us that the whole can exceed the sum of its parts.

Systems thinking in schools is more difficult than in many business settings because of internal segmentation—both horizontal and vertical—that leads many educators to focus on their most immediate settings (Spillane, 1998). Districts are poorly connected to state and national initiatives; relationships within districts are articulated around administrative and reporting requirements, not around educational purpose. The "loosely linked" nature of schools means that activities in one part of the system (finance policy, for example) may have no clear impact on a particular school or classroom. Similarly, important policy decisions in one school (to eliminate tracked curriculum in order to foster more equitable student achievement) may be carried out without affecting other schools in the same administrative unit (district, local education authority). Thus, systems thinking to promote organizational learning needs to be substantially redefined if it is to have practical implications for school-based practitioners.

Mastery—personal and group

Our concept of development also reflects the underlying assumption that organizational learning is lifelong, both for the individual and the team. Senge (1990) emphasizes that: "Personal mastery is the discipline of continually clarifying and deepening our personal vision, of focusing our energies, of developing patience, and of seeing reality objectively. As such, it is an essential cornerstone of the learning organization—the learning organization's spiritual foundation." In school development the reflective teacher who studies and analyzes his or her own teaching roles against the background of specific pupils' needs and research findings exemplifies this ideal. It also means moving beyond individual analysis to make individual observations and data discussible with colleagues: "How am I doing? Why am I doing it this way (are there alternative ways)?" and "Am I doing the right thing?" Louis, Kruse and Raywid (1996) note that "the willingness of teachers to display their craft publicly" (p. 17) was at the heart of the school that they use as an example of a learning organization.

Most of what we know about reflective practice is obtained from studies where it has occurred naturally—in other words, settings in which there has been genuine school development, rather than isolated planned change initiatives. Little has been written about how the school development model could be used to consider interventions in an existing school to promote personal and group mastery.

Mental models

"Mental models" are deeply ingrained assumptions that influence how we understand the world and how we take action. Usually we are not consciously aware of our mental models or the effects they have on our behavior. They are like glasses that we cannot take off and through which we see the world; they are "the way we do things here." The discipline of working with mental models is a hard one, because it involves analyzing aspects of school culture that are implicit. School diagnosis can be a helpful instrument to help schools look inward and make undisclosed assumptions explicit: in other words, *Making Schools Smarter* (Leithwood & Aitken, 1995). This discipline also includes the ability to carry on "learningful" conversations that hold these assumptions up to rigorous scrutiny, that balance inquiry and advocacy, and that expose educators to the influence of others.

Educators have "mental models" of teaching and learning, of being a professional, being a team member, of school leadership, etc. By articulating these models, pieces of the autonomous, evolutionary dimension of organizational life that we embed in daily routine become conscious. When they are, even briefly, open to discussion, the proportions allocated between the developmental dimensions of Figure 11.1 may also temporarily shift, allowing open consideration of the relationship between overt purposes for change, and internal conditions that are rarely considered as part of the equation.

Shared vision

On building a shared vision Senge (1990, p. 9) states: "When there is a genuine vision (as opposed to the all-too-familiar 'vision statement'), people excel and learn, not because they are told to, but because they want to." Louis and Miles (1990) argue that vision emerges from collective reflection on the results of motivated and energetic action, rather than being antecedent to it. This idea is central to the organizational transformation model. However, it also conditions how individuals and teams react to "normal crises." Crisis may reduce team effectiveness if members have a hard time agreeing about, for example, core activities or principles that need to be preserved. On the other hand, effective action as a cohesive team may reinforce a sense of common destiny and commitment. Many leaders in schools—both teachers and administrators—have personal visions that never get translated into shared visions that galvanize their colleagues.

What is lacking is a discipline for translating individual vision into shared vision—not a cookbook but a set of principles and guiding practices. It concerns the third loop of the learning process: the *purpose* or *identity* questions. The practice of shared vision involves the skills of unearthing shared pictures of the future that foster genuine commitment and enrollment rather than compliance (Louis & Miles, 1990, p. 9). Those pictures of the future have to be challenging but realistic and attainable too. In the shared vision, all learning and information (all disciplines) come together as a basis for pictures of the future.

Leveraging random events

Senge's five disciplines of organizational learning directly address our contention that we must understand development as more than planned or managed activities. He also, less obviously, however, addresses our claim that the way in which organizations deal with unanticipated, non-evolutionary stimuli is critical to the development process. He does this through his notion of High leverage, non-obvious solutions to difficult problems. Senge (1990) argues that "learning to see underlying structures [systems] rather than events is a starting point" (p. 65). Louis and Miles (1990) refer to the importance of "coping skills" in working out organizational adjustments to unanticipated opportunities. Coping involves deliberate consideration of random events and threats by looking at the level and intensity of the required reaction, and the systemic implications of the perceived threat or opportunity. The high leverage response may be to do nothing, to make minor adjustments, or to restructure the entire system. In other words, it is the school's ability to incorporate system changes that may, at first, be perceived as highly threatening into the ongoing dialogue about the "how? why? to what end?" questions.

Conclusion

Our argument has been complex, but can be briefly summarized. First, to improve schools we must expand our understanding of the full range of dimensions of events and activities that affect both the primary and secondary organizational processes. Second, we view existing traditions of "planned educational change" as closely related to more recent ideas concerning organizational learning and school development. They provide much of the basis for understanding how and why organizational learning will occur. Third, each tradition of planned educational change described above is limited, and is inadequate both for understanding and stimulated school development and organizational learning. Together, they provide a basis for understanding the "15% direct impact." Finally, the organizational learning and school development frameworks outlined above provide a set of analytic categories (Table 11.2) and core "disciplines" or processes that may extend our ability to make the best use of the available leverage points.

References

Allegro, J.T. (1973). *Sociotechnische organisatie-ontwikkeling*. Leiden: Stenfert Kroese.

Argyris, C. (1990). *Overcoming organizational defenses: Facilitating organizational learning*. Boston: Allyn & Bacon.

Argyris, C. & Schön, D.A. (1978). *Organizational learning: A theory of action perspective*. Reading, MA: Addison-Wesley.

Beare, H. & Boyd, W.L. (eds.) (1993). *Restructuring schools: An international perspective on the movement to transform the control and performance of schools*. Washington/London: Falmer Press.

Beer, M. & Eisenstadt, B. (1990). Why change programs don't produce change. *Harvard Business Review* 68, Nov–Dec: 158–159.

Berg, R.M. van den, Hameyer, U. & Louis, K.S., (eds.) (1987). *Disseminating successful practices*. Leuven, Amersfoort: ACCO.

Berg, R. van den, Sleegers, P., Bakx, E. & Eerden, E. van der (1994). Het innovatief vermogen van scholen in voortgezet onderwijs: Een kwalitatief vooronderzoek. *Pedagogische Studien* 71: 402–419.

Boyd, W.L. & Louis, K.S. (1997). *School development and organizational learning: Towards an integrative theory*. Paper presented at the annual meeting of the American Educational Research Association, Chicago.

Brunsson, N. (1985). *The irrational organization*. New York: Wiley.

Bryk, A., Camburn, E. & Louis, K.S. (1995). Promoting school improvement through professional communities: An analysis of Chicago elementary schools. Paper presented at the annual meeting of the American Educational Research Association, Chicago, IL.

Caldwell, B.J. & Spinks, J.M. (1988). *The self-managing school*. London: Falmer Press.

Caluwé, L. de, Marx, E.C.H. & Petri, M.W. (1988). *School development: Models and change*. Leuven, Amersfoort: ACCO.

Clark, D., Lotto, S. & Astuto, T. (1984). Effective schools and school improvement: A comparative analysis of two lines of inquiry. *Educational Administration Quarterly,* 20 (3): 41– 68.

Cohen, M.D., March, J.G. & Olsen, J.P. (1972). A garbage can model of organizational choice. *Administrative Science Quarterly* 17: 1–25.

Creemers, B.P.M. (1987). Bijdragen tot de effectiviteit van scholen: Discussie. In J. Scheerens & W.G.R. Stoel (ed.), *Effectiviteit van onderwijsorganisaties* (pp. 159–168). Lisse: Swets & Zeitlinger.

Daft, R. & Huber, G. (1987). How organizations learn. In N. DiTomaso & S. Bacharach (eds.), *Research in sociology of organizations,* volume 5. Greenwich: JAI Press.

Deming, W.E. (1988). *Out of the crisis*. Cambridge, MA: MIT Press.

Dimmock, C. (1995). Reconceptualizing restructuring for school effectiveness and school improvement. *International Journal of Educational Reform* 4 (3): 285–300.

Easterby-Smith, M. (1990). Creating a learning organization. *Personnel Review* 19 (5): 24–28.

Edmonds, R. (1979). Effective schools for the urban poor. *Educational Leadership* 39: 15–27.

Fiol, C.M. & Lyles, M.A. (1985). Organizational learning. *Academy of Management Review* 10 (4): 803–813.

Fullan, M.G., Miles, M.B. & Taylor, G. (1980). Organizational development in schools: The state of the art. *Review of Educational Research* 50: 121–183.

Hanson, E.M. (1979). *Educational administration and organization*. Boston: Allyn & Bacon.

Hopkins, D. (1982). Survey feedback as an organization development intervention in educational settings: A review. *Educational Management and Administration* 10: 203–215.

Hopkins, D. (1995). Towards effective school improvement: A comment on the "learning consortium" special issue. *School Effectiveness and School Improvement* 6 (3): 265–274.

Kyle, R.M.J. (ed.) (1985). *Reaching for excellence: An effective schools sourcebook*. Washington, DC: U.S. Government Printing Office.

Lagerweij, N.A.J. & Voogt, J.C. (1990). Policy making at the school level: Some issues for the 90s. *School Effectiveness and School Improvement* 1 (2): 98–120.

Lagerweij, N.A.J. & Voogt, J.C. (1997). *Schoolontwikkeling: Meer dan het invoeren van vernieuwingen alleen.* In Liber Amicorum voor J.G.H.I., Giesbers.

Lawrence, P. & Lorsch, J. (1967). *Organization and environment: Managing differentiation and integration.* Cambridge: Harvard University Press.

Leithwood, K. & Aitken, R. (1995). *Making schools smarter: A system for monitoring school and district progress.* Thousand Oaks, CA: Corwin Press.

Lewin, K. (1947). Frontiers in group dynamics. *Human Relations* 1: 5–41.

Louis, K.S. (1994). Beyond "managed change": Rethinking how schools improve. *School Effectiveness and School Improvement* 5 (1): 2–24.

Louis, K.S., Kruse, S. & Raywid, M. (1996). Putting teachers at the center of reform: Learning schools and professional communities. *NASSP Bulletin* May: 9–21.

Louis, K.S. & Miles, M.B. (1990). *Improving the urban high school: What works and why.* New York: Teachers College Press.

Louis, K.S. & Voogt, J.C. (1997). *Beleidsverzorging verzorgingsbeleid.* Zeist: Voogt.

McDonald, S. (1995). Learning to change: An information perspective on learning in the organization. *Organization Science* 6: 557–568.

Malen, B., Ogawa, R.T. & Kranz, J. (1990). What do we know about school-based management? A case study of literature: A call for research. In W.H. Clune & J.F. Witte (eds.), *Choice and control in American education, vol. 2: The practice of choice, decentralization and school restructuring.* Basingstoke: The Falmer Press.

Marks, H. & Louis, K.S. (1997). Does teacher empowerment affect the classroom? The implications of teacher empowerment for instructional practice and student academic performance. *Educational Evaluation and Policy Quarterly.*

Marx, E.C.H. (1975). *De organisatie van scholengemeenschappen in onderwijskundige optiek.* Groningen: Tjeenk Willink.

Marx, E.C.H. (1987). Meer autonomie in het onderwijs. *Meso* 8 (9): 2–7.

Mintzberg, H. (1979). *The structuring of organizations.* Englewood Cliffs, NJ: Prentice-Hall.

Mintzberg, H. (1994). *The rise and the fall of strategic planning.* New York London: Prentice-Hall.

Morgan, G. (1997). *Images of organizations.* Beverly Hills: Sage Publications.

Morgan, G. & Zohar, A. (1997). Het 15-procent-principe: Geef uw werk een hefboomeffect. *Holland Management Review:* 53.

Murphy, J. (1991). *Restructuring schools: Capturing the phenomena.* New York: Teachers College Press.

Newmann, F. & Wehlege, G. (1995). *Successful school restructuring.* Madison, WI: Center for Effective Secondary Schools, Wisconsin Center for Educational Research.

Owens, R.G. (1981). *Organizational behavior in education.* Englewood Cliffs, NJ: Prentice-Hall.

Pedler, M., Boydell, T. & Burgoyne, J. (1991). *The learning company: A strategy for sustainable development.* London: McGraw-Hill.

Peters, T.J. & Waterman, R.H. (1982). *In search of excellence.* New York: Harper & Row.

Peterson, P.L., McCarthy, S.J. & Elmore, R.F. (1996). Learning from school restructuring. *American Educational Research Journal* 33 (1): 119–153.

Purkey, S. & Smith, M. (1983) Effective schools: A review. *Elementary School Journal,* 83, 427–452.

Quinn, R.E. & Rohrbaugh, J. (1983). A spatial model of effectiveness criteria: Towards a competing values approach to organizational analysis. *Management Science* 29 (3): 363–377.

Resnick, L., Nolan, K. & Resnick, D. (1995). Benchmarking education standards. *Educational Evaluation and Policy Analysis* 17 (4): 438–461.

Reynolds, D., Bollen, R., Creemers, B.P.M., Hopkins, D., Stoll, L. & Lagerweij, N.A.J. (1996). *Making good schools: Linking school effectiveness and school improvement.* London/New York: Routledge.

Rollow, S. & Bryk, A. (1995). Catalyzing professional community in a school reform left behind. In K.S. Louis & S. Kruse (eds.), *Professionalism and community: Perspectives on reforming urban schools.* Thousand Oaks, CA: Corwin.

Scheerens, J. (1987). (invited address) *Het zelfevaluerend vermogen van onderwijs-instellingen.* Enschede: Universiteit van Twente.

Scheerens, J. & Creemers, B.P.M. (1989). Towards a more comprehensive conceptualization of school effectiveness. In B. Creemers, T. Peters & D. Reynolds (eds.), *School effectiveness and school improvement* (pp. 265–278). Lisse, Amsterdam, Rockland, MA, Berwyn, PA: Swets & Zeitlinger.

Schein, E.H. (1992). *Organizational culture and leadership.* San Francisco London: Jossey-Bass.

Schmuck, R.A., & Runkel, P.J. (1985). *The Handbook of organizational development in schools* (third edition). Palo Alto, CA.

Schmuck, R.A., Runkel, P.J., Arends, J.H. & Arends, R.I. (1977). *The second handbook of organizational development in schools.* Palo Alto, CA.

Senge, P.M. (1990). *The fifth discipline.* New York: Doubleday.

Sergiovanni, T. (1994). *Building community in schools.* San Francisco: Jossey-Bass.

Smylie, M., Lazarus, V. & Brownlee-Conyers, J. (1996). Instructional outcomes of school-based participative decision-making. *Educational Evaluation and Policy Analysis* 18 (3): 181–198.

Spillane, J. (1998). State policy and the non-monolithic nature of the local school district: Organizational and professional considerations. *American Educational Research Journal* 35 (1): 33–63.

Sproull, L., Weiner, S. & Wolf, D. (1978). *Organizing an anarchy: Belief, bureaucracy and politics in the National Institute of Education.* Chicago: University of Chicago Press.

Swieringa, J. & Wierdsma, A.F.M. (1990). *Op weg naar een lerende organisatie.* Groningen: Wolters-Noordhoff.

Thomas, S., Sammons, P., Mortimore, P. & Smees, R. (1995). *Stability and consistency in secondary schools' effects on students' GCSE outcomes over three years.* Paper presented at the annual meeting of the International Congress of School Effectiveness and School Improvement, Leeuwarden, NL.

Tjepkema, S. (1994). *Profiel van de lerende organisatie en haar opleidingsfunctie.* Enschede: Universiteit Twente.

U.S. Department of Education (1986). *What works: Research about teaching and learning.* Washington, DC: USDE.

Vandenberghe, R. (1987). *De schoolbetrokken zelfanalyse als voorbereiding op het V.L.O. project.* Beschrijving van activiteiten in twaalf S.B.A.-scholen voor en na hun toetreding tot het V.L.O.-project. Vierde rapportage. Leuven: Kath. Universiteit, Centrum voor Onderwijsbeleid en Vernieuwing.

Van Velzen, W.G., Miles, M.B., Ekholm, M. & Robin, D. (1985). *Making school improvement work.* Leuven/Amersfoort: ACCO.

Van Wieringen, A.M.L. (1993). Transformatie naar zelfscholende onderwijsinstellingen. In P.J.J. Stijnen, J. Scheerens, A.M.L. Van Wieringen, H.G.W. Münstermann (eds.), *Transformatie van schoolorganisaties* (p. 302). Alphen aan de Rijn: Samson Tjeenk Willink.

Voogt, J.C. (1986). Werken aan onderwijsverbetering. In J.C. Voogt & A. Reints (ed.), *Naar beter onderwijs* (pp. 101–125). Tilburg: Zwijsen.

Voogt, J.C. (1989). *Scholen doorgelicht: Een studie over schooldiagnose*. Proefschrift. De Lier: ABC.

Voogt, J.C. (1995). *Innovatiebeleid, flankerend beleid en parallel beleid: Pleidooi voor consistent beleid*. Den Haag: PMB (interne notitie).

Voogt, J.C., Louis, K.S. & Van Wieringen, A.M.L. (1997). *Deregulation and decentralization in the Netherlands: The case of the educational services system*. Paper presented at the annual meeting of the American Educational Research Association, Chicago.

Weiler, H. (1993). Comparative perspectives on educational decentralization: An exercise in contradiction? *Educational Evaluation and Policy Analysis* 12: 443–448.

Willms, J.D. & Raudenbusch, S. (1989). A longitudinal hierarchical linear model for estimating school effects and their stability. *Journal of Educational Measurement* 26 (3): 209–232.

Part IV

New Directions

12

Organizational Learning and Current Reform Efforts: From Exploitation to Exploration[1]

Sam Stringfield

Introduction

Much of the evidence about organizational learning is limited to what James March (1996) refers to as "exploitative" forms of such learning. Exploitative learning entails "such things as refinement, choice, production, efficiency, selection, implementation, execution" (p. 102). Exploitative learning builds incrementally on the skills and knowledge already possessed by the organization and its members. In the short run, this is a productive form of learning for the organization. It is "cashing in" on the investments already made in developing effective and efficient organizational responses. In the long run, however, an organization that limits itself to only this form of learning stands a good chance of becoming obsolete. Challenges likely to be encountered by schools and many other types of organizations are not likely to be resolved by incremental learning alone. Such challenges require exploratory learning as well, learning which considers radical departures from existing practices. "The problem," as March argues, "of balancing exploration and exploitation is exhibited in distinctions made between refinement of an existing technology and invention of a new one . . . exploration of new alternatives reduces the speed with which skills at existing ones are improved" (p. 102).

1 This chapter was written under funding from the Office of Educational Research and Improvement, U.S. Department of Education (No. R-117-d-40005). However, any opinions expressed do not necessarily represent the positions or policies of the U.S. Department of Education.

Problems associated with the need for balance between exploitative and exploratory forms of organizational learning manifest themselves in schools as trade-offs between local school improvement efforts which aim to refine existing practices, as distinct from adopting and learning how to use dramatically new practices. In this chapter, I explore a new generation of educational innovations and argue that adopting such innovations, while costly in the short run, is a better alternative in a significant number of schools than is local school improvement, which is largely dependent on exploitative forms of organizational learning.

The meaning of "innovation"

The book most associated with the phrase "learning organizations" is Peter Senge's (1990) *The fifth discipline: The art and practice of the learning organization*. That thoughtfully crafted volume begins with an analogy of one of the great successes of the twentieth century: air travel. Senge described the Wright Brothers' 1903 first flight as an "invention," in that it proved that human flight in a heavier-than-air craft was possible. Senge noted that, in engineering, an invention becomes an "innovation" only when it can be replicated on a meaningful scale at practical costs (pp. 5–6). He elaborated that the Douglas Corporation's DC-3, which went on the market in 1935, was the first "innovation" in air travel. Senge was careful to detail that by "innovation" he meant that the DC-3 was the first reliable, commercially practical craft.

The DC-3 involved five new "component technologies." Used in ensemble, those five new technologies allowed the Douglas Corporation to launch the huge set of industries that today include aircraft design and construction, airlines, airports, and hundreds of related service industries worldwide. Senge described how only one year before the DC-3, the Boeing corporation had produced a new airplane that the company's leadership had hoped would revolutionize air transportation. The design used only four of the five "component technologies," and proved unstable on take-off and landing. Boeing's engineers had given it their best shot, but their design was deficient by one component, and the plane was never a commercial success. Though innovative, the Boeing 247 was not an "innovation."

Senge used the DC-3 story as an analogy, and a segue into his discussion of what he believed are the five component parts, or "disciplines," of a learning organization: systems thinking, personal mastery, mental models, building shared vision, and team learning. However, the "idea, invention, innovation" distinction, and the importance of (an initially unknown number of) component technologies necessary to move an idea through invention to innovation are amply worthy of consideration on their own merits. As Senge (1990) stated, "At their best, efforts to develop learning capabilities blend 'behavioral' and 'technical' changes" (p. xvii).

This chapter builds on the assumption that Senge's basic metaphor and limiting definitions should be taken more seriously in education. We should be much more conservative in how we label ideas for school improvement. Most of what are termed "innovative programs" in education have rarely or never been proven to produce gains in students' achievements. They are simply ideas that, in the eyes of their proponents, are lovely. For example, however attractive I and others find the majority of the New American Schools designs (see Stringfield, Ross & Smith, 1996), until each of the designs clearly demonstrates that it has improved student achievement in at least a small number of schools (a task on which each NAS design team is currently working diligently), each should be described as an "idea."[2]

Some educational reform designs have demonstrated that they have produced desired student outcomes in a few environments, and those could be regarded, in Senge's typology, as "inventions." Ellis and Fouts (1997), Herman and Stringfield (1997), and Fashola and Slavin (1997) all review substantial subsets of the research on outcomes associated with various reforms, and these reviews have produced at least partially overlapping lists of programs and school reform designs that, at least in some circumstances, have been demonstrated to "work."

An example of an "invention" that apparently will not be adequately tested to determine whether or not it could be regarded as an "innovation" is the Calvert School Curriculum. Calvert School is a century-old private school in Baltimore, Maryland. The board of directors of Calvert allowed the school's highly-articulated curriculum and very detailed instructional program to be implemented in two high-poverty elementary schools in Baltimore City School System. Evaluations of each effort (McHugh & Spath, 1997; Stringfield, 1995) indicated gains in student's academic achievements. Across two-to-four years of careful observation and the administration of multiple tests measuring reading, writing/language arts, and mathematics, researchers noted that cohorts of students receiving the Calvert program consistently produced mean achievement scores that were 20–30 percentile points higher than pre-Calvert students in the same neighborhood schools. Yet the board of directors of the private Calvert school simply has declined to further disseminate their copyrighted whole-school materials and design.

Even though Calvert's board did not choose to replicate their success beyond two public schools, it is useful, when considering the necessary components of other programs that do wish to scale-up, to consider Calvert's key features. Those key features include the following:

1. *A more stimulating, more demanding, more fully articulated curriculum for all students.* The curriculum is worked out in great detail, and modified slightly every year-to-few-years, based on the experiences of Calvert's teachers and actively-involved administration. Teachers and administrators at Barclay and Woodson were encouraged to make similarly context-sensitive alterations, but the basic curriculum and curriculum philosophy were to remain intact.

---

2 I am one of many who have ignored this important distinction.

2. *Writing every day.* Calvert requires that students write about their read-ings, their art, their history lessons. Writing begins in first grade with sim-ple sentences, but by fourth grade has evolved into weekly, multi-page, 100 percent grammatically correct discourse.

3. *A full-time program coordinator.* From the beginning, the Calvert admin-istration insisted that both Barclay and Woodson maintain a full-time pro-gram coordinator. Principals already have full-time jobs. The Calvert program requires changes from traditional modes in every hour of a teacher's day. For such a sea-change to succeed, Calvert insisted that each school have a full-time Calvert coordinator.

4. *Initial, and ongoing, focused staff development.* This begins with two weeks of summer preparation time, and includes both whole-faculty and individual-teacher development throughout the year. The Calvert coordi-nator often models lessons in teachers' classes, and provides teachers with feedback on their subsequent lessons.

5. *Folders of students' works, regularly checked.* At the back of every Calvert, Barclay, and Woodson class, there are hanging folders of each student's work. Each day the folders are checked, and needed corrections noted, by the teacher or aide. Students' first 15 minutes of the next day are spent making corrections or in silent reading. Over time, students learn that it is easier to do assignments carefully and correctly than to have to make repeated corrections. Each month the coordinator (or other mem-bers of the faculty or administration) checks each student's folder for con-tent and accuracy, then the folder is sent home to the child's parent(s) or guardian(s). This process provides unambiguous content for parent–teacher and teacher–administration conversations about a stu-dent's progress. Are there specific things parents could be working on with their child at home? Is the teacher giving specific feedback on the student's work? Is the administration making sure that the class has the supports it needs to keep moving forward through the curriculum? A great deal of specificity in discourse is gained through the use of regularly-checked stu-dent folders.

6. *Making sure that things work.* This is easy to overlook and hard to over-estimate. It is common in urban, public education for a school to experi-ence recurring shortages of such simple materials as paper and pencils. Often whole grades don't get a specific new text until well into the school year. In the Calvert program, all of Barclay's and Woodson's teachers receive a full year's supply of texts and supporting materials before the school year begins. If a class comes close to running out of, for example, pencils, additional pencils are ordered immediately. Pencils are very inex-pensive. Downtime in schools is expensive. Interestingly within this area, it was after Barclay's teachers first visited Calvert's large materials stor-age room, and were assured that they could have year-long access to its contents, that they immediately, overwhelmingly voted to implement the program.

Within Senge's definitions, many proposals for school restructuring could never become "innovations." A few, like Calvert, choose not to reproduce their successes. Many others focus on a goal of each school reforming itself in its own way. The idea of unique reforms for each school faces several serious problems. One is intellectual. In his 1899 presidential address before the American Psychological Association, John Dewey[3] described the sheer implausibility of expecting teachers and principals, who were fully employed in their professional obligations, to also create meaningful school reforms. Rather, Dewey called for the development of a "linking science" between practice and theory. Dewey argued that such group of reform developers and modifiers could assist practitioners in avoiding feeling compelled to:

> fall back upon mere routine traditions of school teaching, or fly to the latest fad of pedagogical theorists—the latest panacea peddled out in school journals or teachers' institutes—just as the old physician relied upon his magic formulas. (p. 113)

The second significant problem with infinitely repeating unique inventions as a route to improving the academic performance of students is that the large-scale studies have indicated that the majority of such restructuring efforts have not demonstrated positive student effects (Crandall & Loucks, 1983, Stringfield et al., 1997; and for the special case of movement to site-based management, Murphy & Beck, 1995). The majority of these efforts, some quite well funded, have not demonstrated that they can be considered "inventions," let alone "innovations."

The third problem has been where the "reforming each school uniquely" line of reasoning has been that, even when successful, locally developed reforms can be remarkably expensive. Crandall & Loucks (1983) found that the per-teacher cost of local development was often "20 times the [cost of] adoption" (p. 22) of an externally developed design. Given that education is perpetually underfunded, an emphasis on site-by-site development of school improvement strategies may represent a luxury beyond the reach of most schools and school systems.

Applying Senge's definitions to education today, a distinctly smaller set of reform ideas and designs could be described as "innovations." Four examples from this gradually growing field follow. Reviews of research on the High/Scope preschool program (Schweinhart et al., 1986; Weikart et al., 1971) have indicated that the program can be effective, and it has been disseminated in all 50 states. More than one type of cooperative learning (e.g. Stevens et al., 1987; Stevens & Slavin, 1995) has been found effective at improving students' academic achievements, and cooperative learning techniques have been widely replicated. Direct Instruction (DI) is an example of an elementary school program that has been widely implemented and evaluated, and generally found to produce positive student outcomes (Becker & Gersten, 1982).

---

3 Dewey's address was published in the *Psychological Review* the following year.

Success for All/Roots and Wings (Slavin et al., 1996a, 1996b) is an example of a whole school restructuring approach that has produced positive student outcomes on a variety of measures in diverse contexts. As with Calvert, some of the particulars of Success for All (SFA) are worthy of note, and are described below.

1. *The SFA reading curriculum is based on research on effective reading instruction in beginning reading, and on effective uses of cooperative learning.*

2. *The progress of all students is assessed every eight weeks.* Short SFA reading tests are enhanced with teachers' and the program facilitator's professional judgements, and students' assignments are modified accordingly.

3. *Reading tutors.* Because a core objective of SFA is to detect and correct reading problems before they become engrained, and because one-to-one tutoring is perhaps the most research-proven intervention in education, SFA invests in reading tutors for first graders who are in danger of falling behind.

4. *Preschool and kindergarten.* To the extent that local or federal Title I funds can support, SFA invests in full-day preschool and academic kindergarten programs. This is consistent with the developers' philosophy of prevention and early intervention.

5. *Family support team.* Parents and guardians are the people who logically can provide the most individualized attention and other supports to a child. Yet parents often face substantial hurdles of their own. The role of the Family support team is to assist families in obtaining the services they need, and to thereby help families stay focused on the needs of their children.

6. *Program facilitator.* Researchers have gathered data on the components of the program without which SFA simply will not succeed. An active, involved program facilitator is very high on the SFA team's list of non-negotiables. SFA's experience in hundreds of high-poverty schools is that principals are largely tied down with organizational and bureaucratic tasks. A full-time SFA facilitator becomes critical.

7. *Teacher training.* Teachers receive detailed teachers' manuals to support the detailed reading materials. Teachers and aides receive three full days of training before beginning SFA, and additional training throughout each year. In addition, SFA operates regional SFA conferences, so that leadership teams can continuously upgrade their skills. As with Calvert, the facilitator models lessons in the classes of teachers who are finding particular units challenging.

8. *Advisory committee.* The building principal, program facilitator, teacher representatives, parent representatives, and family support staff meet regularly to review the school's and program's progress, and to solve emerging problems. In addition, grade-level teams meet regularly.

9. *Special education.* To the greatest extent possible SFA strives to "never-stream" students. Through early identification of problems, and through

concentrating scarce resources in the early grades, SFA attempts to keep children in their regular classrooms throughout their school days, and years.

10. *Relentlessness.* An over-arching attitude of SFA is relentlessness. So long as one child is in danger of falling behind, there is specific work that needs doing. Combined with a carefully designed, research-based curricular and instructional program, relentlessness is the condition necessary to ensure that no child "falls through the cracks."

An important note of caution is required. Several of the same studies that have demonstrated the sometimes-remarkable advantages of some types of thoughtful school reform have also documented the occasional failure of those same reform efforts in improving student achievements. For example, in the Special Strategies Studies (Stringfield et al., 1997), one Success for All school obtained truly remarkable three-year gains in students' reading levels for whole cohorts of initially low-achieving students. Supporting qualitative data documented the extent to which the school had gone at initial implementation with gusto, and had sustained their focus and enthusiasm through nearly ten years of normal teacher movement and the loss of their initial SFA-supporting principal. The new principal had been initially indifferent to SFA, but the program's support among the faculty, combined with the new principal's eventual observations of the program at work and the aggressive re-intervention of SFA trainers from Johns Hopkins, resulted in a sustained high level of implementation.

Less than two hours away, a second inner-city school observed during *Special Strategies* obtained very mildly positive results using SFA. Why? Three years of detailed observation indicated that when this school lost its SFA-supporting principal, the replacement person was never well-socialized into SFA, and never supported the program. Many teachers drifted back to their old and familiar methods, new teachers were not provided with initial or ongoing training, and the program never regained momentum, or full implementation.

Similar Special Strategies data were obtained at two Comer School Development Program sites: strong implementation leads to strong student results, weak implementation to weak student results. In both cases the data supported the hypothesis that the programs were potentially valid, but that cross-site implementation had not been highly reliable.

Needed: High reliability reform

Gathering longitudinal data for the Louisiana School Effectiveness study (Stringfield & Teddlie, 1991), I was repeatedly struck that the studies "positive outlier" (higher achieving) schools were not necessarily "trendy." Rather, they were places where almost all of the important things worked almost all of the time. This condition was readily contrasted with the "negative outlier" schools, which were easily characterized as places where a great many potentially important things were regularly allowed to fail. The effect of these regularly recurring, often severe, within-school, specific and general system failures was that teachers learned to retreat to the safety of their own classrooms. Over time, the schools might or might not take on various new ideas for school improvement, but these reforms rarely worked and never lasted. Even if the new reforms were valid in the abstract, the specific levels of implementation were so unreliable at so many levels that (often repeated) efforts at school improvement failed.

Part of what seemed odd was that the same improvement effort that succeeded in a more effective school (measured as a dimension controlled for community socio-economic status) would invariably fail in a less effective school. Clearly, one of the first rules of research, that reliability sets the upper boundary of validity, was also working in educational practice.

While there is very little research on the reliability of schools and school reform efforts, there is a growing body of knowledge on the necessary conditions for very high reliability of human performance in other fields. In the remainder of this section, major components of that knowledge base are described, with references to the two cases detailed above, school reform generally, and Senge's five disciplines of learning organizations.

Regardless of the field, the condition under which high reliability organizations (HROs) evolve is that the public and the professionals working within an organization come to believe that failure of the organization to achieve the organization's core goals would be disastrous. For example, it was after the public came to perceive high costs associated with mid-air collisions of planes that modern, extraordinarily reliable air-traffic control was born.

From the end of World War II until the early 1970s, average workers' incomes rose steeply, regardless of their level of education. However, from 1973 until 1995, the median income of young male high-school dropouts in this country has dropped by over 50% in constant dollars (Murnane & Levy, 1996). When almost every family's economic health was on a long upswing, failures of educational reform efforts were intellectually interesting, but otherwise unimportant. However, since the mid-1970s, the practical costs of failed reforms and failing students have risen dramatically. America's policy community is groping for what Britain's Prime Minister, Tony Blair, calls "zero-tolerance for school failure." Level of education is a better predictor of a person's economic future than at any time since such data were first collected. The issues today are practical, not theoretical. The first condition for the evolution of schools and school systems into HROs, that the public and professionals within the field come to view failure as potentially disastrous, is increasingly being met.

In the cases of the Calvert School replications at Barclay and Woodson, and the Success for All dissemination efforts across the country, both sets of developers/disseminators regard the success of their dissemination efforts as ideologically and personally critical. This becomes part of the "relentlessness" of SFA, and explains the hours volunteered by the headmaster of Calvert, whenever Barclay or Woodson experienced problems.

In both Calvert cases, and in the successful SFA and Comer cases from Special Strategies, the principals were possessed of a fiery determination that this reform would succeed for these children, now. All of the principals had previously tried other reforms with less success. All were committed to success this time with this reform.

HROs require clarity regarding goals. No person or organization can sustain remarkable levels of reliability in a huge diversity of areas. Each HRO has a strong sense of its primary mission. Within Senge's (1990) conception of learning organizations, shared vision is one of five disciplines. The result is the same: the evolution of a few, common, tightly-held goals is central to an HRO's reliability.

In both SFA/Roots and Wings, and Calvert, students' academic achievements in the core areas of reading, writing, and mathematics consistently come before a variety of other organizational concerns. In both, the institutional goal is that all children succeed in their academic lives. Social goals are valued, but, in both programs, faculty discussions are not allowed to drift far from individual students' routes to academic success. In the relatively unsuccessful SFA case in Special Strategies, the new principal was more committed to her ideas than to measures of student success. The far from spectacular results were almost preordained. Fashola and Slavin's (1997) review of common features of more effective programs found that the first common characteristic was clear goals.

HROs recruit staff aggressively, and then train and retrain constantly. High levels of organizational reliability are the result of systemic human effort. Senge discusses team learning and personal mastery as key to learning organizations. HROs formalize both disciplines. From either perspective, a major theme is that no system can overcome poorly trained, unmotivated, inattentive members of the core professional team.

The Calvert headmaster and staff actively sought, nominated, and encouraged highly qualified teachers to apply for positions at Barclay and Woodson. Similarly, the Barclay and Woodson principals used their affiliation with Calvert as a drawing card when seeking new teachers. Within Special Strategies, the highly successful SFA school's faculty was actively involved in the choice of a new principal. By contrast, at the low implementation school, a new principal was simply "assigned from the central office."

Fashola and Slavin found that clearly targeted, content-rich, long-term professional development was a second cross-program characteristic of effective reform efforts. A separate chapter in this volume addresses the importance of carefully targeted professional development in creating schools as learning communities.

HROs take performance evaluations seriously. Highly reliable organizations must constantly rely on the professional judgements of all of their team members. Egregious failures of professional judgement can not coexist with even moderate-term reliability.

One of the advantages that accrue to a principal operating a school with a more fully articulated academic program is that she can more quickly and clearly see who is or is not currently able or willing to teach the curriculum. The principal of Woodson remarked that the first two years of the Calvert program exposed dozens of presumably long-standing weaknesses among her faculty. The majority of these problems could be addressed through staff development, but the principal also counseled one teacher into retirement and another into a different choice of profession. Similar stories can be found in many SFA schools.

In HROs, monitoring is mutual without counterproductive loss of overall autonomy and confidence. In air-traffic control towers, it becomes critical that everyone knows what those around them are doing. Because the team perceives the importance of their task, and because a single person's failure can almost instantly place the lives of hundreds of airline passengers at grave risk, monitoring of one another's strengths and limitations is continuous. Supervisors often take on front-line responsibilities for a few minutes or hours, and the tower staff engage in a rich web of formal and informal feedback. By focusing on getting the job right, every single time, the goal of enhanced team professionalism remains clear throughout often rather pointed feedback.

In both the Calvert program and SFA, facilitators model desirable techniques more than lecture about them. They provide feedback, often critical feedback, but in a climate of focus on student success, not undifferentiated personnel evaluation. Typically, the effect is, in Senge's terminology, team learning.

HROs are alert to surprises or lapses. Everyone makes mistakes. Surprises come every day. Some of them are delightful, and others court disaster. Highly reliable organizations cannot eliminate surprises or human lapses. Rather, HRO staffs become skilled at focusing on the types of small problems that can rapidly cascade into large problems.

The concept of avoiding cascading errors is, in learning organizations language, a systems-thinking skill. Cascades take time, and often spill across traditional lines of authority. Diverting potentially cascading errors is necessarily a combination of a systems challenge and a challenge to personal mastery.

In most high-poverty, urban school systems that gather annual testing information, poor students do not merely begin behind their more advantaged peers, they fall further behind each year. It is not at all uncommon in urban elementary schools to meet third and fourth grade students who read at first-grade level. Often these students are perfectly intelligent, and one sign of their intelligence is that they have surmised, or learned, that they cannot learn to read. Efforts to teach these fourth graders to read are greatly complicated by the students' knowledge that they cannot learn to read. A problem that at many SFA schools and at Woodson was caught early has instead been allowed to cascade through years, and by grade four will be very hard to reverse. No small part of the strength of both SFA and Calvert is their relentless determination to catch lapses before they can cascade into students' learned perception of inability to learn.

HROs build powerful databases on dimensions highly relevant to the organizations' core goals. Those databases feature triangulating data on central dimensions and real-time availability. The databases are regularly cross-checked by multiple concerned groups. Close, ongoing attention to students' progress was another common feature of programs found to be more effective in the Fashola and Slavin (1997) review.

SFA's combination of quarterly testing, teachers' judgements, and facilitators' observations work together to create rich, relevant, regularly used and cross-checked information to all concerned. Nine times each year, the Calvert folders provide parents, teachers, and administrators with a clear picture of each student's strengths, areas in need of improvement, and rate of progress.

HROs extend formal, logical decision analysis as far as extant knowledge will allow. Time is the enemy of real-world professional judgement. Cognitive psychology has made clear that humans deal with rapidly overlapping demands by establishing mental routines and "automaticity" in recurring areas. The evolution of standard operating procedures (SOPs) in some areas can free up a professional's mind to apply sophisticated judgement in novel areas.

Professional development can help teams respond in similar events in similar ways. It can provide a shared vocabulary for team discussions of unusual events. SFA, Calvert, and virtually all of the other programs that Ellis and Fouts (1997), Herman and Stringfield (1997), and Fashola and Slavin (1997) have identified as more effective share the characteristic of having standardized significant portions of the working of the school and classroom. These are not efforts to bound teachers' creativity, but rather are the provision of mutually-understood forms in which teachers can express and share their professional skills and insights.

HROs have initiatives that identify flaws in standard operating procedures. They nominate and validate changes in their own procedures, whenever old procedures have proven inadequate. HROs resolve the tension between a rapidly changing world and the need for some standardization in some areas by constantly honoring those that identify flaws in "what is" and evolve better models of "what we could be."

From the beginning of the experiment, the Barclay and Woodson faculties have been encouraged to find and address mismatches between the Calvert curriculum and their students. SFA is currently on its "third generation" reading program. By continuously gathering and analyzing data on the inevitable modifications individual schools and districts make in the program, the SFA research team is able to better understand "what works, where, and why?" and make appropriate modifications.

HROs are hierarchically structured, but, during times of peak load, HROs display a second layer of behavior that emphasizes collegial decision-making, regardless of rank. This second mode is characterized by cooperation and coordination. During times of peak activity, line staff are expected to exercise considerable professional discretion. Relationships are complex, coupled, and often urgent.

When a group engages in systems thinking, building shared vision, and team learning, that group is almost necessarily entering into a world where relationships are often fluid and generally dynamic. One of the lessons of HROs is that such dynamism is not incompatible with hierarchy. In fact, both become necessary. Hierarchy provides efficiency during periods of normal workload. But in times of peak load, the very structure of the system is designed to give authority to the professional (regardless of formal rank) who is most involved in the immediate task.

Several of the current school restructuring designs deliberately build rich webs of professional relationships among traditionally isolated teachers and administrators. These webs can have the effect of increasing the entire team's confidence in—and, during busy times, trust in the ceding of authority to— diverse members of the team. It is entirely possible that much of educational reform has focused too much on whether the static condition of educational organizations is hierarchical or flat. The more important question may not be "whether," but "when."

HROs are invariably valued by their supervising organizations. The activities that create unusual levels of reliability must be understood and protected by the higher levels of the system. Disinterest is not a condition that generates understanding or protection. One of the advantages shared by both the "Title I Program Improvement" requirements and the targeting of schools for improvement (or state take-over) in many countries' current "high-stakes testing" schemes is that the efforts are focusing more attention on low-performing schools than has been so focused in years—probably ever. Monitoring can be mutual only if all levels are monitoring, and, in HROs, diverse levels monitor one another.

Returning to the two examples detailed early in this chapter, both SFA and Calvert early on institutionalized systems of feedback to their participating schools. SFA now annually obtains written feedback from third-party site monitors, and asks all participating teachers to rate each professional development session. The teachers' ratings are returned to the SFA central office, not the day's presenter. The monitoring hasn't always been agreeable, but its presence has helped keep staff focused on clear goals, and has often guided future professional development activities.

Summary

Building from Dewey's (1900) call for a "linking science" between educational theory and practice, several major studies have found that the intelligent importation of educational "innovations" (e.g. programs proven to achieve their stated goals in multiple previous settings, and to be replicable at a knowable cost) is both possible and, in many circumstances, a preferable route to reform. Such a route entails, as I suggested at the outset, a form of exploratory learning not common in schools, because of its short-term costs. This chapter has considered common themes between those reforms, the conditions and steps necessary to reliably implement them (contextualized exploratory learning processes) and the relationships between promising designs, high reliability, and learning organizations.

In *The Fifth Discipline*, Senge (1990) argued that, "At their best, efforts to develop learning capabilities blend 'behavioral' and 'technical' changes" (p. xvii). In both the book's introduction and "coda," Senge uses the DC-3 as an example of more technical change. This chapter has focused on the technology of our educational research base in support of education's development of our first DC-3s. We are slowly evolving better designs for student achievement. As we develop more nearly valid designs, we, and the worried, taxpaying public, are pushing toward increasing the reliability of traditional schools, and of reform implementations.

It is entirely possible that these developments in educational ideas, inventions, and innovations can produce gains in student achievement in the twenty-first century that are every bit as dramatic as our twentieth-century gains in sophistication regarding heavier-than-air flight. At that point, our ancestors can take the same smug amusement at our "state of the art" educational reforms that we take when viewing the Wright brothers' plane and the DC-3, assuming schools can arrive at a more productive balance between exploitative and exploratory forms of collective learning. We can all look forward to future organizations that have learned to be Boeing 777s of schools.

References

Becker, W.C. & Gersten, R. (1982). A follow-up of Follow Through: The later effects of the Direct Instruction model on children in fifth and sixth grades. *American Educational Research Journal* 19 (1): 75–92.

Crandall, D.P. & Loucks, S.F. (1983). *A roadmap for school improvement: Executive summary of the Study of Dissemination Efforts Supporting School Improvement.* Andover, MA: The NETWORK.

Dewey, J. (1900). Psychology and social practice. *The Psychological Review* 7 (2): 105–124.

Ellis, A.K. & Fouts, J.T. (1997). *Research on educational innovations.* Larchmont, NY: Eye on Education.

Fashola, T. & Slavin, R. (1997). Promising programs for elementary and middle schools: Evidence of effectiveness and replicability. *Journal of Education for Students Placed At Risk* 2 (3): 251–308.

Herman, R. & Stringfield, S. (1997). *Ten promising programs for educating all students: Evidence of impact.* Arlington, VA: Educational Research Service.

McHugh, B. & Spath, S. (1997). Carter G. Woodson Elementary School: The success of a private school curriculum in an urban public school. *Journal of Education for Students Placed at Risk* 2 (2): 121–136.

March, J. (1996). Exploration and exploitation in organizational learning. In M. Cohen & L.S. Sproull (eds.), *Organizational learning*. Thousand Oaks, CA: Sage Publications.

Murnane, R.J. & Levy, F. (1996). *Teaching the new basic skills: Principles for educating children to thrive in a changing economy*. New York: The Free Press.

Murphy, J. & Beck, L. (1995). *School-based management as school reform: Taking stock*. Newbury Park, CA: Corwin.

Schweinhart, L.J., Weikart, D.P. & Larner, M.B. (1986). Consequences of three preschool curriculum models through age 15. *Early Childhood Research Quarterly*, 1: 15–45.

Senge, P.M. (1990). *The fifth discipline*. New York: Doubleday.

Slavin, R.E., Madden, N.A., Dolan, L.J., Wasik, B.A., Ross, S., Smith, L. & Dianda, M. (1996a). Success for all: A summary of research. *Journal of Education for Students Placed at Risk* 1: 41–76.

Slavin, R.E., Madden, N.A. & Wasik, B.A. (1996b). Roots and wings. In S. Stringfield, S. Ross & L. Smith (eds.), *Bold plans for school restructuring: The New American Schools Development Corporation designs* (pp. 207–231). Mahwah, NJ: Erlbaum.

Stevens, R.J., Madden, N.A., Slavin, R.E. & Farnish, A.M. (1987). Cooperative integrated reading and composition: Two field experiments. *Reading Research Quarterly* 22: 433–454.

Stevens, R.J. & Slavin, R.E. (1995). Effects of a cooperative approach in reading and writing on academically handicapped and nonhandicapped students. *The Elementary School Journal* 95: 241–262.

Stringfield, S. (1995). *Fourth-year evaluation of the Barclay/Calvert experiment*. Baltimore: Johns Hopkins University Center for the Social Organization of Schools.

Stringfield, S., Millsap, M., Herman, R., Yoder, N., Brigham, N., Nesselrodt, P., Schaffer, E., Karweit, N., Levin, M. & Stevens, R. (1997). *Special strategies studies final report*. Washington, DC: U.S. Department of Education.

Stringfield, S., Ross, S. & Smith, L., (eds.) (1996). *Bold plans for school restructuring: The New American Schools Development Corporation designs*. Mahwah, NJ: Erlbaum.

Stringfield, S. & Teddlie, C. (1991). Observers as predictors of schools' multi-year outlier status. *Elementary School Journal* 91 (4): 357–376.

Weikart, D.P., Rogers, L., Adcock, C. & McClelland, D. (1971). *The cognitively oriented curriculum: A framework for preschool teachers*. Urbana: University of Illinois.

13

From Organizational Learning to Professional Learning Communities

Karen Seashore Louis and Kenneth Leithwood

> Organizing and learning are essentially antithetical processes, which means the phrase "organizational learning" qualifies as an oxymoron. To learn is to disorganize and increase variety. To organize is to forget and reduce variety. (Weick & Westley, 1996)

Introduction

This book has been about organizational learning processes in schools. The intent of the book was to fill gaps that are left by the increasing popularity of the concept among educational scholars and practitioners, and the relatively limited empirical research in school settings. Thus, our emphasis has been largely on action: images of organizational learning embedded in school contexts, interventions for enhancing organizational learning processes, the outcomes of such processes, and the relationship between organizational learning and the processes of school improvement. By focusing on the unique contexts of school organizations, the chapters have also provided evidence about how concepts of organizational learning apply to public educational settings. Finally, although previous literature concerning organizational learning processes describes extensively both what it is in real organizations and what it ought to be under ideal conditions, there are few sources of defensible advice about strategies to actually enhance it. This book described a large number of such strategies, four of which are given chapter-length attention, including systematic evidence of their effects.

The chapters have helped us develop a good sense of *what* school members do and should do in the interests of their own individual and collective learning but have said very little about *why* they would want to. In other words, except for Chapter 2 (Louis & Kruse) and Chapter 3 (Ben-Peretz & Schonmann) we have not explicitly addressed the organizational contexts in which members of a school would collectively commit themselves to continuous professional learning and improvement using the processes that have been described.

We conclude this volume by considering the conditions under which school members would be likely to commit themselves to continuous, collective, professional learning. We, like Weick and Westley (1996), argue that the disequilibrium introduced by organizational learning must be balanced by features of the school organization that support equilibrium:

- "Productive" organizational learning—the kind that pays off for students—occurs when school members consistently take collective responsibility for student learning.
- Collective responsibility depends on the school organization having stable, community-like characteristics, especially for the staff.
- The conditions which foster organizational learning overlap in significant degree with the conditions that create a sense of community among staff members.
- An image or vision of schools as *professional learning communities* holds considerable promise as a response to the challenges facing education now and in the near future, as Linda Darling-Hammond characterized them in the opening paragraphs of Chapter 1.

Organizational Learning and disequilibrium in schools

Dictionary definitions of learning provide us with a wide variety of familiar but inconsistent images. To learn is to be socialized, which implies a strong connection with traditional values and ways of doing things. It is also to memorize, which implies a stable and receptive capacity. On the other hand, learning is also very active: comprehending, grasping, mastering.

Some of the images of organizational learning that we have encountered in most of the empirical chapters in this book are dynamic and even dramatic—in the extreme—for example, the Okanagon school described in Louis and Kruse (Chapter 2) is "self-designing" in that there is a continuous emphasis on restructuring and adjusting. The "honoring of differences" emerged as an important theme in Mitchell and Sackney's discussion of how schools come to common conceptions of their work. In other words, the more commonality that was achieved, the more diversity was encouraged. Leithwood, Jantzi, and Steinbach

(Chapter 4) note that principals in learning schools encourage varied learning strategies and opportunities rather than systemic activities that are always aligned with core goals. Ben-Peretz and Schonmann (Chapter 3) emphasize the drama and unpredictability of learning in teachers' lounges: the "learning moment" often occurs during a mini-crisis of confidence and performance. While the dramatic moments are not chronicled in Darling-Hammond, Cobb and Bullmaster's contribution (Chapter 7), they re-emphasize the importance of educational processes as highly problematic, demanding non-routine experimentation in order to energize teachers and create commitment.

The discontinuous and unpredictable nature of learning schools is perhaps most poignantly described in King, Allen, and Nguyen (Chapter 5), where teachers who had been most involved in the school's reform agenda over a period of five years expressed copious discontent about the lack of cohesiveness in the school's change processes and decision-making. In place of what they experienced, they longed for "just common sense, good cooperative learning for adults . . . when you really want 'em to talk and give 'em a task to do, pros and cons, and come out with that and then come back to the whole group and do a quick report." In fact, one of the major lessons that King et al. draw is that the reality of change (individual maps), even in a small school that had been collectively involved in a change effort over a long period, were more complex than the group could deal with.

One of the dilemmas that is posed by the theme of learning as disequilibrium is that the more organizational learning has occurred in schools, the more likely people are to recognize continuing differences and the need for continuing change. As Weick (1979) points out, doubt becomes an institutionalized feature of the school, encouraging collective and voiced dissatisfaction with the status quo rather than individual and undiscussed disquiet.

Disequilibrium and "learning from small failures and wins"

Disequilibrium is a necessary part of any transformative process. We support the notion that transformational change is important in today's schools; nothing less is likely to address the significant shifts in demographic, economic, and social conditions that form the relevant environment for schools (Reyes & Wagstaff, in press). In addition, as Leithwood (1994) and Leithwood et al. (1995) have shown, transformative leadership is critical to creating school environments that are focused on professional expertise and are open to change.

At the same time, we argue that the emphasis on transformation and re-invention must be balanced by the idea of evolutionary and developmental changes emphasized in Voogt et al. (Chapter 11), and the "connectionist" perspective on group learning that is described by Leithwood (Chapter 9). Both of these chapters revisit the distinction between "single-loop learning" and "double-loop learning" (Argyris & Schön, 1978), or knowledge that is applied within a culture contrasted with knowledge that changes a culture (see Cousins, Chapter 10). Much of the organizational learning literature assumes that double-loop/culture change learning is, by definition, superior and more difficult. Our chapters and our experience in other change settings affirm that both are problematic and important in schools, and that they need to be nurtured. Thus, the theoretical chapters in section III, as well as the empirical chapters, suggest that the dramatic learning moments in schools are balanced with more continuous and organized efforts to learn that may contribute to improved student achievement without requiring transformation.

A concept of learning as a series of "small wins" is consistent with other new directions in organizational change research (Morgan, 1997). The life cycle metaphor, for example, has described organization on the basis of birth, maturity, and death (Kimberly & Miles, 1981). Organizational ecology has used the Darwinist analogy of the "survival of the fittest," and has examined the mimetic character of successful organizations—in other words, their tendency over time to look increasingly alike (Hannen & Freeman, 1989). Contingency theories, on the other hand, have generated a metaphor of situation-specific adjustment between organizational characteristics and environmental "learning" situations (Miles, 1980). Most life-cycle theorists assume that, in the absence of dramatic changes in the environment, organizations will be best served by adaptive change processes. Others, such as Weick (1974), have argued that educational organizations are unusual systems, in that they are "loosely coupled," a characteristic that makes change all the less likely to occur rapidly, or to affect the whole organization in dramatic ways. Furthermore, we point to the fact that collective or organizational learning is, under many circumstances, a fragmented and disconnected process. If we look, for example, at Cousin's description of the learning that accompanied participatory evaluation (Chapter six), we see ample evidence of the importance of unanticipated small crises that generated different forms of learning than the designers of a formal change initiative anticipated. We also see that sub-groups learned from the same events at different levels and in different ways. There is little question in the analysis that the "event" of participatory evaluation did not produce immediate or uniform learning in a school or school system. In addition, the "spread" of learning was limited to those groups that collectively experienced discussions about the program and the evaluation. Also, there was considerable evidence of differential within-team learning among groups and individuals that were only loosely connected in conversation.

King et al.'s analysis of Thomas Paine High School suggests, furthermore, that there is a temporal dissociation between deep learning and action that may limit the value of formal "learning interventions" under some circumstances. The authors present substantial evidence that there was progress among all participants in grasping fundamental characteristics of the school, its processes, and its effects. On the other hand, King et al. remain concerned that uncovering and understanding key issues related to the school did not necessarily make concerted, collective action any easier. While we can make the assumption that what was learned will affect future behavior, King is skeptical that fundamental, double-loop organizational learning and short-term problem-solving are intimately connected—a perspective that is more consistent with evolutionary than transformational change theories. Nevertheless, although the "cognitive mapping" exercise had little direct impact on Thomas Paine High School's immediate needs, the process revealed and deepened the staff's sense of common understanding of the dynamics of the school's reform efforts.

In order to reflect on how and when dynamic forms of organizational learning may result in effective and positive change, we paradoxically turn to the source of organizational stability in schools: professional community.

Community in schools and the need for stability and commitment

A community, according to Bryk and Driscoll (1988), "is a social organization consisting of cooperative relations among adults who share common purposes and where daily life for both adults and students is organized in ways which foster commitment among its members." As Selznick (1992) further explains, its function "is to regulate, discipline, and especially to channel self-regarding conduct, thereby binding it so far as is possible, to comprehensive interests and ideals," and its favored form "is the small, intimate, person-centered structure where solidarity is most effective and most genuine . . . where persons are created and nurtured, where they become situated beings and implicated selves" (p. 369). In other words, the notions of stable patterns of trust, mutual interdependence, and permanent personal investment to the group are core. Sergiovanni (1994), like other theorists who advocate community, argues that affiliation and commitment to the group are basic human needs. He also points out that philosophers like Nel Noddings who also advocate caring and commitment emphasize that it is important because it guarantees *continuity and stability* in relationships and routines (p. 51). In other words, unlike the image of disequilibrium engendered by organizational learning, the image that a school community stimulates is one of dependable memberships, reliable practices, consensual norms, and explicit structures.

The idea of *professional community* builds on the community metaphor by incorporating a strong emphasis on the professionalization of teachers' work through increasing expert knowledge. Recent investigations of teachers' work point to many ways in which teachers use collegial resources, scholarly and action research as resources for professional refreshment. Professional community may be promoted by educational networks or other professional organizations beyond the school. A school's involvement in discipline-based discussions, such as those surrounding the development of state and national standards, can become the basis for teacher commitment and interaction, particularly in high schools (McLaughlin, 1993; Rowan, 1994; Siskin, 1991). But the type of professional community that is embedded in schools, where teachers come together around meaningful, shared issues irrespective of their disciplinary base is even more powerful (Sergiovanni, 1994). The development of school-wide community does not mean that other forms of collective professional relationships are unimportant, nor that they are incompatible with departmental structures or disciplinary networks.

Making major changes in day-to-day classroom work without the support and pressure of peers engaged in complementary efforts is possible but very difficult for teachers. Professional community is important, not only because it supports but because it reinforces peer pressure and accountability on staff who may not have carried their fair share, as well as easing the burden on teachers who have worked hard in isolation but who have felt unable to help some students. Thus, by combining social and intellectual support, professional community may lead to improved practices that can increase student achievement and other forms of school effectiveness (Louis et al., 1996).

School-based professional community

Strong school-based professional communities do not ensure stable, positive organizational cultures, but they may promote it through five critical elements (Kruse et al., 1995).

Shared norms and values

Members of a school may develop and reaffirm, through language and action, their common assumptions about children, learning, teaching and teachers' roles, the importance of interpersonal connection, and commitment to the collective good (Bryk et al., 1993). No one set of values is necessary, but they typically address teachers' views about children and their ability to learn, priorities for the use of time and space within a school setting, and the proper roles of parents, teachers and administrators. They assume that all students can learn at reasonably high levels, and that teachers can help them, despite many obstacles that students may face outside of school.

Focus on student learning

Concentration on student learning is a core characteristic of professional communities (Sergiovanni, 1994). Teachers' professional discussion and action centers on improving students' opportunity to learn, and seeks to enhance student benefit (Abbott, 1991; Darling-Hammond & Goodwin, 1993). Teachers talk about ways in which their actions promote students' intellectual development, as distinguished from simply focusing on activities or techniques that may engage student attention. Within a strong professional community, this focus is enforced by mutual obligations among teachers rather than by rules. When teachers send clear and consistent messages to one another about the objectives and methods of learning, learning is more likely, because student and faculty effort can be directed more effectively toward intellectual ends.

De-privatization of practice

Reflective dialogue is often coupled with increasingly open discussion about individual teachers' practice. Teachers who are moving toward authentic pedagogy must, of course, rely heavily on their own insights and analysis of how effective they are in stimulating students to create and skillfully manipulate knowledge. However, because there are no formulas about how to do this, peers become a critical source of insight and feedback to improve personal understanding of one's own practice. Thus, where the norms of autonomy are diminished, teachers within an emerging professional community become committed to practicing their craft in public ways (Lieberman, 1988). Achieving authentic pedagogy requires continuous teacher judgement about forms of instruction that are not well defined in textbooks or practice, such as how to balance depth of coverage of important concepts with breadth, and how to elicit reflective work from students in classrooms. By sharing uncertainties about practice, teachers learn new ways to talk about what they do, and kindle new relationships between participants that focus on their craft.

Collaboration
Collaboration is a natural outgrowth of reflective dialogue and deprivatized practice in which teachers move beyond mutual understanding to the production of materials and activities that improve instruction, curriculum and assessment practices for students, and new approaches to staff development for themselves. As teachers work with students from increasingly diverse social backgrounds, and as the curriculum begins to demand more intellectual rigor, teachers require information and increased technical competence that can be enhanced through sharing their expertise. Collaborative efforts also augment the mosaic of socio-emotional support within the school (Little, 1990). When teachers collaborate productively, they participate in reflective dialogue, they observe and react to one another's teaching, curriculum and assessment practices, and they engage in joint planning and curriculum development—all of which improve practice.

Reflective dialogue
Reflection promotes teachers' awareness of their practice and its consequences. Commitment to reflection as a communal activity means regular conversation among teachers focusing on the academic, curricular, and instructional concerns of practice within the school, as well as on student development (Zeichner & Tabachnick, 1991). Members of the community can use these discussions to evaluate themselves as well as the institution within which they work.

Intersections between organizational learning and community

At first glance, the emphasis of organizational learning theory on disequilibrium and that of professional community on commitments that create a consensus about goals and methods would seem incompatible. However, recent research, including the chapters in this book, implies that they emerge, in fact, under the same set of organizational conditions. Here we will point to just a few that are visible in the previous chapters.

Louis and Kruse (Chapter 2) point to a number of structural similarities between two schools characterized by high levels of organizational learning:

time for teachers to meet and work together, high levels of interdependence among teachers, and teacher empowerment. Louis, Marks, and Kruse (1996) demonstrated a strong statistical association between these same structural features of 24 schools, and their levels of professional community, while Bryk, Camburn, and Louis (forthcoming) show that teacher involvement in school governance activities and committees is a strong predictor of both professional community and organizational learning in Chicago elementary schools. Similarly, Smylie (1996) equates teacher learning with human capital development, and regards it as a vehicle for school reform.

Leithwood, Jantzi, and Steinbach (Chapter 4) make a persuasive argument that collaborative and collegial school cultures (elements of professional community) and strong leadership are key factors in stimulating organizational learning. Bryk, Camburn, and Louis, using a different implicit model, show that leadership that is simultaneously authoritative and supportive is strongly associated with professional community, and that professional community is, in turn, a strong predictor of organizational learning. A similar emphasis on the relationship between leadership behaviors and professional community is made in Louis et al. (1996).

Finally, trust is viewed by many as a precondition for risk-taking behaviors that may lead to organizational learning of a more fundamental sort (Hargreaves, 1991; Myerson et al., 1996). Without trusting relationships in schools, teachers may be afraid that even small losses are not worth efforts to arrange their classrooms differently, or to change the status quo in the school, even if it is acknowledged that current practices are not very effective. At the same time, trust is viewed as a prerequisite for community: Bryk, Lee and Holland (1993) argue, for example, that one essential difference between the Catholic and public school sectors is the establishment of trust between parents and teachers. This trust permits teachers to engage in more caring relationships with their students.

Empirical investigations of the relationship between organizational learning and professional community are, at this juncture, very limited, and assertions of causality may be misplaced. Yet, there is reason to argue that certain conditions in the school seem to be associated both with the development of cohesive and trusting relationships among teachers in schools—*stable professional community*—and also with active efforts to engage with new ideas in ways that challenge current practice—*team and organizational learning*. At the very least, we are drawn to the conclusion that the apparent tension between stability and change, between community and learning, is a positive dynamic.

From organizational learning to professional learning communities: the role of collective responsibility

When school values are vague or when consensus on what is expected of teachers is low, teachers may feel comfortable with the autonomy they have to pursue their unique interests. But individual autonomy can reduce teacher efficacy when teachers cannot count on colleagues to reinforce their objectives. In contrast, clear shared values and norms maximize teacher success through collective rein-forcement. Values are reflected in practices: all teachers might require students who are failing to take part in after-school support activities, which demonstrates that they consistently appreciate student achievement, and take responsibility for giving extra help to ensure that learning occurs. But "learning" can occur with-in a stable, unquestioned but flawed paradigm, just as injustice and even cor-ruption can flourish in groups where there are high levels of internal trust and consensus. This, for example, is the lesson of *The Shopping Mall High School* (Powell et al., 1985). The appearance of effective schooling can occur in settings where there is an evolutionary collusion between teachers, and among teachers and students, to demand less and less productivity in return for superficially orderly behavior—what Senge (1990) would call a learning disability.

What occurs in these situations of organizational learning failures within diminished communities is a deflection of goals to a focus on meeting the imme-diate needs of both students and teachers for a peaceful, pleasant working envi-ronment. We have increasing evidence that collective responsibility for student learning, in addition to improved technical teaching practices and curriculum, is a fundamental correlate of student achievement (Newmann et al., 1996). Unless professional communities are consistently focused on this objective, and unless learning opportunities—whether small or transformational—are focused on this objective, the potential for entropy is great.

Concluding remarks

Let us return to the paradox with which we began. Schools as organizations contain both a need for change and disorderliness, and also a demand for sta-bility and a cohesive story about "where we have been and where we are going." Schools need to engage in fundamental and risky learning, and they need to organize themselves into communities of caring and trust. These two processes must coexist, for the chapters presented in this volume suggest that they are interdependent as well as oppositional. As Weick and Westley (1997) point out:

> The likelihood of learning drops quickly when invention and disor-der overwhelm capacities for retention and identify, or when systems, routines, and order overwhelm capacities for unjustified variation. These tendencies towards overwhelming are a constant threat because each one represents a simpler way of dealing with the world. (p. 456)

Schools, like other organizations, must live with this complexity.

References

Abbott, A. (1991). The order of professionalization: An empirical analysis. *Work and Occupations* 18 (4): 355–384.

Argyris, C. & Schön, D. (1978). *Organizational Learning*. Reading, MA: Addison-Wesley.

Bryk, A.S., Camburn, E. & Louis, K.S. (forthcoming). Professional community in Chicago elementary schools.

Bryk, A.S. & Driscoll, M. (1988). *An empirical investigation of the school as community*. University of Chicago: Unpublished manuscript.

Bryk, A.S., Lee, V. & Holland, P. (1993). *Catholic schools and the common good*. Cambridge, MA: Harvard University Press.

Darling-Hammond, L. & Goodwin, A.L. (1993). Progress toward professionalism in teaching. In G. Cawelti (ed.), *Challenges and achievements of American education: The ASCD 1993 yearbook*. Alexandria, VA: Association for Supervision and Curriculum Development.

Hannen, M. & Freeman, H. (1989). *Organizational ecology*. Cambridge, MA: Harvard University Press.

Hargreaves, A. (1991). *Restructuring restructuring: Postmodernity and the prospects for educational change*. Toronto: Ontario Institute for Studies in Education, University of Toronto.

Kimberly, J. & Miles, R. (1981) *The organizational life cycle: Issues in the creation, transformation and decline of organizations*. San Francisco: Jossey-Bass.

Leithwood, K. (1994) Leadership for school restructuring. *Educational Administration Quarterly 30* (4): 498–518.

Leithwood, K., Jantzi, D. & Steinbach, R. (1995). An organizational learning perspective on schools' responses to external policy initiatives. *School Organization 15* (3): 229–252.

Lieberman, A. (1988). *Building a professional culture in schools*. New York: Teachers College Press.

Little, J.W. (1990). The persistence of privacy: Autonomy and initiative in teachers' professional relations. *Teachers College Record 91* (4): 509–536.

Louis, K.S., Marks, H. & Kruse, S. (1996). Teachers' professional community in restructuring schools. *American Educational Research Journal 33*(4): 757–798.

McLaughlin, M.W. (1993). *What matters most in teachers' workplace context?* Washington, DC: Office of Educational Research and Improvement.

Miles, R.H. (1980). *Macro organizational behavior*. Santa Monica: Goodyear.

Myerson, D., Weick, K. & Kramer, R. (1996). Swift trust and temporary groups. In R. Kramer & T. Tyler (eds.), *Trust in organizations* (pp. 166–195). Thousand Oaks: Sage Publications.

Morgan, G. (1997). *Images of organization* (second ed.). Thousand Oaks, CA: Sage Publications.

Newmann, F. & associates (1996). *Authentic achievement: Restructuring schools for intellectual quality*. San Francisco: Jossey-Bass.

Powell, A., Farrar, E. & Cohen, D. (1985). *The shopping mall high school: Winners and losers in the educational marketplace*. Boston: Houghton-Mifflin.

Reyes, P. & Wagstaff, L. (in press). Delta forces: The changing fabric of American society and education. In J. Murphy & K.S. Louis (eds.), *Handbook of Educational Administration* (second edition). San Francisco: Jossey Bass.

Rowan, B. (1994). Comparing teachers' work with other occupations: Notes on the professional status of teaching. *Educational Researcher 23*: 4–18.

Selznick, P. (1992). *The moral commonwealth: Social theory and the promise of community*. Berkeley, CA: University of California Press.

Sergiovanni, T.J. (1994). *Building community in schools*. San Francisco: Jossey-Bass.

Siskin, L. (1991). Departments as different worlds: Subject subcultures in secondary schools. *Educational Administration Quarterly* 27: 134–160.

Smylie, M.A. (1996). From bureaucratic control to building human capital: The importance of teacher learning in education reform. *Educational Researcher* 25 (9): 9–11.

Weick, K. (1979). *The social psychology of organizing* (second edition). Reading, MA: Addison-Wesley.

Weick, K. & Westley, F. (1996). Organizational learning: Affirming an oxymoron. In S. Clegg, C. Hardy & W. Nord (eds.), *Handbook of organizational studies*. Thousand Oaks: Sage Publications.

Zeichner, K.M. & Tabachnick, B.R. (1991). Reflections on reflective teaching. In B.R. Tabachnick & K.M. Zeichner (eds.), *Issues and practices in inquiry oriented teacher education* (pp. 1–21). London: Falmer Press.

Author Index

Subject Index

Milton Keynes UK
Ingram Content Group UK Ltd.
UKHW031145141024
449569UK00024B/1066

9 789026 515408